T0184026

Lecture Notes
in Business Information Processing   348

Series Editors

Wil van der Aalst
   RWTH Aachen University, Aachen, Germany
John Mylopoulos
   University of Trento, Trento, Italy
Michael Rosemann
   Queensland University of Technology, Brisbane, QLD, Australia
Michael J. Shaw
   University of Illinois, Urbana-Champaign, IL, USA
Clemens Szyperski
   Microsoft Research, Redmond, WA, USA

More information about this series at http://www.springer.com/series/7911

Paulo Sérgio Abreu Freitas ·
Fatima Dargam · José Maria Moreno (Eds.)

Decision Support Systems IX

Main Developments and Future Trends

5th International Conference on Decision
Support System Technology, EmC-ICDSST 2019
Funchal, Madeira, Portugal, May 27–29, 2019
Proceedings

 Springer

Editors
Paulo Sérgio Abreu Freitas
Universidade da Madeira
Funchal, Madeira, Portugal

Fatima Dargam
SimTech Simulation Technology
Graz, Austria

José Maria Moreno
University of Zaragoza
Zaragoza, Spain

ISSN 1865-1348 ISSN 1865-1356 (electronic)
Lecture Notes in Business Information Processing
ISBN 978-3-030-18818-4 ISBN 978-3-030-18819-1 (eBook)
https://doi.org/10.1007/978-3-030-18819-1

This Springer imprint is published by the registered company Springer Nature Switzerland AG
The registered company address is: Gewerbestrasse 11, 6330 Cham, Switzerland

EURO Working Group on Decision Support Systems

The EWG-DSS is a Euro Working Group on Decision Support Systems within EURO, the Association of the European Operational Research Societies. The main purpose of the EWG-DSS is to establish a platform for encouraging state-of-the-art high-quality research and collaboration work within the DSS community. Other aims of the EWG-DSS are to:

- Encourage the exchange of information among practitioners, end-users, and researchers in the area of decision systems
- Enforce the networking among the DSS communities available and facilitate activities that are essential for the start up of international cooperation research and projects
- Facilitate the creation of professional, academic, and industrial opportunities for its members
- Favor the development of innovative models, methods, and tools in the field of decision support and related areas
- Actively promote the interest on decision systems in the scientific community by organizing dedicated workshops, seminars, mini-conferences, and conference, as well as editing special and contributed issues in relevant scientific journals

The EWG-DSS was founded with 24 members, during the EURO Summer Institute on DSS that took place at Madeira, Portugal, in May 1989, organized by two well- known academics of the OR community: Jean-Pierre Brans and José Paixão. The EWG-DSS group has substantially grown along the years. Currently, we have over 350 registered members from around the world.

Through the years, much collaboration among the group members has generated valuable contributions to the DSS field, which resulted in many journal publications. Since its creation, the EWG-DSS has held annual meetings in various European countries, and has taken active part in the EURO Conferences on decision-making-related subjects. Starting from 2015, the EWG-DSS established its own annual conferences, namely, the International Conference on Decision Support System Technology (ICDSST).

The current EWG-DSS Coordination Board comprises of seven experienced scholars and practitioners in the DSS field: Pascale Zaraté (France), Fátima Dargam (Austria), Shaofeng Liu (UK), Boris Delibašic (Serbia), Isabelle Linden (Belgium), Jason Papathanasiou (Greece) and Pavlos Delias (Greece).

Preface

The proceedings of the ninth edition of the EWG-DSS Decision Support Systems published in the LNBIP series present a selection of reviewed and revised full papers from the EURO Mini Conference and 5th International Conference on Decision Support System Technology (EmC-ICDSST 2019) held in Madeira, Portugal, during May 27–29, 2019, with the main theme: "Decision Support Systems: Main Developments and Future Trends." This event was jointly organized by the EURO Association of European Operational Research Societies and the EURO Working Group on Decision Support Systems (EWG-DSS) and it was hosted by the University of Madeira (UMA) in Funchal, Portugal.

The EWG-DSS series of the International Conference on Decision Support System Technology (ICDSST), starting with ICDSST 2015, was planned to consolidate the tradition of annual events organized by the EWG-DSS in offering a platform for European and international DSS communities, comprising the academic and industrial sectors, to present state-of-the-art DSS research and developments, to discuss current challenges that surround decision-making processes, to exchange ideas about realistic and innovative solutions, and to co-develop potential business opportunities. This year ICDSST 2019 was organized as EURO Mini-Conference (EmC-ICDSST 2019) and had the theme of "DSS: Main Developments and Future Trends" in order to take the opportunity of the celebration of the "EWG-DSS 30th Anniversary" for the conference to evaluate how the research area in DSS has substantially advanced within the past 30 years and how the EWG-DSS has helped the DSS communities to consolidate research and development in the co-related areas, considering its research initiatives and activities.

EmC-ICDSST 2019 recapitulated the developments of the decision support systems area in the past 30 years, enforcing the trends and new technologies in use, so that a consensus about the appropriate steps to be taken in future DSS research work can be established.

The scientific topic areas of EmC-ICDSST 2019 included:

- Advances in research on decision-making and related areas
- Artificial intelligence applied to decision support systems
- Advances in applied decision support systems
- Trends for new developments in decision support systems
- Decision making integrated solutions within open data platforms
- Knowledge management and resource discovery for decision-making
- Decision-making methods, technologies, and real-industry applications
- Geographic information systems and decision-making/support
- Decision-making, knowledge management, and business intelligence
- DSS for business sustainability, innovation, and entrepreneurship
- Decision-making in high and medium education
- Innovative decision-making approaches/methods and technologies

- Big data analytics approaches for solving decision-making issues
- Big data visualization to support decision analysis and decision-making
- Social-networks analysis for decision-making
- Group and collaborative decision-making
- Multi-attribute and multi-criteria decision-making
- Approaches and advances in group decision and negotiation DSS
- Decision support systems and decision-making in the health sector

The aforementioned topics reflect some of the essential topics of decision support systems, and they represent several topics of the research interests of the group members. This rich variety of themes, advertised not only to the (more than 300) members of the group, but to a broader audience as well, allowed us to gather several contributions regarding the implementation of decision support processes, methods, and technologies in a large variety of domains. Hence, this EWG-DSS LNBIP Springer edition has considered contributions of a "full-paper" format, selected through a single-blind paper reviewing process. In particular, at least three reviewers – members of the Program Committee – reviewed each submission through a rigorous two-staged process. Finally, we selected 11 out of 59 submissions, corresponding to a 19% rate, to be included in this 9th EWG-DSS Springer LNBIP edition.

We proudly present the selected contributions, organized in three sections:

1. *Decision Support Systems in Societal Issues:* Cases where decision support can have an impact on society are presented through real-world situations. First, Maria Drakaki, Hacer Güner Gören, and Panagiotis Tzionas use data obtained from management reports for refugees and migrant sites to forecast emotions and potential tensions in local communities. Ana Paula Henriques de Gusmão, Rafaella Maria Aragão Pereira, and Maisa Silva collected georeferenced data from the platform Onde Fui Roubado and the location of the military units of Recife to determine efficient spatial distributions for the police units. Floating taxi data are fed into advanced spatiotemporal dynamic identification techniques to gain a deep understanding of complex relations among urban road paths in the work of Glykeria Myrovali, Theodoros Karakasidis, Avraam Charakopoulos, Panagiotis Tzenos, Maria Morfoulaki, and Georgia Aifadopoulou. This section closes with the work of Guoqing Zhao, Shaofeng Liu, Huilan Chen, Carmen Lopez, Lynne Butel, Jorge Hernandez, Cécile Guyon, Rina Iannacone, Nicola Calabrese, Hervé Panetto, Janusz Kacprzyk, and Mme Alemany on the identification of the causes of food waste generation and of food waste prevention strategies, a critical symptom of modern societies.
2. *Decision Support Systems in Industrial and Business Applications:* Approaches that illustrate the value of decision support in a business context are presented. Herwig Zeiner, Wolfgang Weiss, Roland Unterberger, Dietmar Maurer, and Robert Jöbstl explain how time-aware knowledge graphs can enable us to do time series analysis, discover temporal dependencies between events, and implement time-sensitive applications. George Tsakalidis, Kostas Vergidis, Pavlos Delias, and Maro Vlachopoulou present their approach on how to systematize business processes through a conceptual entity applicable to BPM practices and compliance-checking via a contextual business process structure that sets the boundaries of business

process as a clearly defined entity. This section finishes with the work of Pascale Zaraté, Mme Alemany, Mariana Del Pino, Ana Esteso Alavarez, and Guy Camilleri, who use a group decision support system to help farmers in fixing the price of their production considering several parameters such as harvesting, seeds, ground, season, etc.

3. *Advances in Decision Support Systems Methods and Technologies:* This section highlights methods, techniques, approaches, and technologies that advance the research of the DSS field. Sarra Bouzayane and Inès Saad use a supervised learning technique that allows one to extract the preferences of decision-makers for the action categorization for an incremental periodic prediction problem. Georgios Tsaples, Jason Papathanasiou, Andreas C. Georgiou, and Nikolaos Samaras use a two-stage data envelopment analysis to calculate a sustainability index for 28 European countries. Alejandro Fernandez, Gabriela Bosetti, Sergio Firmenich, and Pascale Zarate present a technology that brings multiple-criteria decision support on Web pages that customers typically visit to make buying decisions. Advances in decision support continue with the work of Oussama Raboun, Eric Chojnacki, and Alexis Tsoukias, who focus on the rating problem and approach it with a novel technique based on an evolving set of profiles characterizing the predefined ordered classes.

We would like to thank all the people who contributed to the production process of this LNBIP book. First of all, we would like to thank Springer for continuously providing EWG-DSS with the opportunity to guest edit the DSS book. We particularly wish to express our sincere gratitude to Ralf Gerstner and Christine Reiss, for their dedication in guiding us during the editing process. Secondly, we thank all the authors for submitting their state-of-the-art work for consideration to this volume, which marks an anniversary of 30 years of the EWG-DSS and confirms to all of us that the DSS community continues to be as active as ever with a great potential for contributions. This encourages and stimulates us to continue the series of International Conferences on DSS Technology. Finally, we express our deep gratitude to all reviewers, members of the Program Committee, who assisted on a volunteer basis in the improvement and the selection of the papers, under the given competitive scenario of the papers and the tight schedule. We believe that the current EWG-DSS Springer LNBIP volume brings together a rigorous selection of high-quality papers addressing various points of decision support systems developments and trends, within the conference theme. We sincerely hope that the readers enjoy the publication!

March 2019   Paulo Sérgio Abreu Freitas
                                                                    Fatima Dargam
                                                                José Maria Moreno

Organization

Conference Organizing Chairs and Local Organizing Team

Paulo Sérgio Abreu Freitas	University of Madeira, Portugal
	paulo.freitas@staff.uma.pt
Rita Ribeiro	UNINOVA University, Lisbon, Portugal
	rar@uninova.pt
Fátima Dargam	SimTech Simulation Technology, Austria
	f.dargam@simtechnology.com
Ana Luisa Respício	University of Lisbon, Portugal
	respicio@di.fc.ul.pt
António Rodrigues	University of Lisbon, Portugal
	ajrodrigues@fc.ul.pt
Jorge Freire de Souza	University of Porto, Portugal
	jfsousa@fe.up.pt
Jorge Nélio Ferreira	University of Madeira, Portugal
	jorge.nelio.ferreira@staff.uma.pt

Program Committee

Adiel Teixeira de Almeida	Federal University of Pernambuco, Brazil
Alex Duffy	University of Strathclyde, UK
Alexander Smirnov	Russian Academy of Sciences, Russia
Alexis Tsoukias	University Paris Dauphine, France
Alok Choudhary	Loughborough University, UK
Ana Paula Cabral	Federal University of Pernambuco, Brazil
Ana Respício	University of Lisbon, Portugal
Andy Wong	University of Strathclyde, UK
Bertrand Mareschal	Université Libre de Bruxelles, Belgium
Boris Delibašić	University of Belgrade, Serbia
Carlos Henggeler Antunes	University of Coimbra, Portugal
Daouda Kamissoko	University of Toulouse, France
Dragana Bečejski-Vujaklija	Serbian Society for Informatics, Serbia
Fátima Dargam	SimTech Simulation Technology/ILTC, Austria
Francisco Antunes	Beira Interior University, Portugal
François Pinet	Cemagref/Irstea, France
Frantisek Sudzina	Aalborg University, Denmark
Gabriela Florescu	National Institute for Research and Development in Informatics, Romania
Guy Camilleri	Toulouse III University/IRIT, France
Hing Kai Chan	University of Nottingham, Ningbo Campus, UK/China
Irène Abi-Zeid	FSA – Laval University, Canada

Isabelle Linden	University of Namur, Belgium
Jan Mares	University of Chemical Technology, Czech Republic
Jason Papathanasiou	University of Macedonia, Greece
Jean-Marie Jacquet	University of Namur, Belgium
João Lourenço	Universidade de Lisboa, Portugal
João Paulo Costa	University of Coimbra, Portugal
Jorge Freire de Souza	Engineering University of Porto, Portugal
José Maria Moreno Jimenez	Zaragoza University, Spain
Kathrin Kirchner	Berlin School of Economics and Law, Germany
Lai Xu	Bournemouth University, UK
Marko Bohanec	Jozef Stefan Institute, Slovenia
Michail Madas	University of Macedonia, Greece
Milosz Kadzinski	Poznan University of Technology, Poland
Nikolaos Matsatsinis	Technical University of Crete, Greece
Nikolaos Ploskas	University of Macedonia, Greece
Panagiota Digkoglou	University of Macedonia, Greece
Pascale Zaraté	IRIT/Toulouse University, France
Pavlos Delias	Eastern Macedonia and Thrace Institute of Technology, Greece
Rita Ribeiro	UNINOVA – CA3, Portugal
Rudolf Vetschera	University of Vienna, Austria
Sandro Radovanović	University of Belgrade, Serbia
Shaofeng Liu	University of Plymouth, UK
Stefanos Tsiaras	Aristotle University of Thessaloniki, Greece
Theodore Tarnanidis	University of Macedonia, Greece
Uchitha Jayawickrama	Staffordshire University, UK
Wim Vanhoof	University of Namur, Belgium
Xin James He	Fairfield University, USA

Steering Committee – EWG-DSS Coordination Board

Pascale Zaraté	IRIT/Toulouse 1 Capitole University, France
Fátima Dargam	SimTech Simulation Technology, Austria
Shaofeng Liu	University of Plymouth, UK
Boris Delibašić	University of Belgrade, Serbia
Isabelle Linden	University of Namur, Belgium
Jason Papathanasiou	University of Macedonia, Greece
Pavlos Delias	Eastern Macedonia and Thrace Institute of Technology, Greece

Sponsors

EURO Working Group on Decision Support Systems
(https://ewgdss.wordpress.com)

EURO - Association of European Operational Research
Societies (www.euro-online.org)

APDIO - Associação Portuguesa de Investigação
Operacional (http://apdio.pt)

Institutional Sponsors

UMA Universidade da Madeira, Portugal
(http://www.uma.pt/)

UNINOVA – CA3 – Computational Intelligence Research
Group Portugal (www.uninova.pt/ca3/)

University of Toulouse, France (http://www.univ-tlse1.fr/)

IRIT Institut de Research en Informatique de Toulouse,
France (http://www.irit.fr/)

SimTech Simulation Technology, Austria
(http://www.SimTechnology.com)

Graduate School of Management, Faculty of Business,
University of Plymouth, UK
(http://www.plymouth.ac.uk/)

Faculty of Organisational Sciences, University of Belgrade, Serbia (http://www.fon.bg.ac.rs/eng/)

University of Namur, Belgium (http://www.unamur.be/)

University of Macedonia, Department of Business Administration, Thessaloniki, Greece (http://www.uom.gr/index.php?newlang=eng)

Eastern Macedonia and Thrace Institute of Technology, Greece (http://www.teiemt.gr)

Industrial Sponsors

REACH Innovation EU (http://www.reach-consultancy.eu/)

Deloitte (https://www2.deloitte.com/)

Springer (www.springer.com)

Lumina Decision Systems (www.lumina.com)

ExtendSim Power Tools for Simulation (http://www.extendsim.com)

Paramount Decisions (https://paramountdecisions.com/)

1000 Minds (https://www.1000minds.com/)

Contents

Advances in Decision Support Systems' Methods and Technologies

Decision Support Systems
in Societal Issues

Fuzzy Cognitive Maps as a Tool to Forecast Emotions in Refugee and Migrant Communities for Site Management

Maria Drakaki[1(✉)], Hacer Güner Gören[2], and Panagiotis Tzionas[1]

[1] Department of Automation Engineering,
Alexander Technological Educational Institute of Thessaloniki,
P.O. Box 141, 574 00 Thessaloniki, Greece
{drakaki,ptzionas}@autom.teithe.gr
[2] Department of Industrial Engineering, Pamukkale University,
Kinikli Campus, Denizli, Turkey
hgoren@pau.edu.tr

Abstract. Refugees and migrants arrivals in the Mediterranean since 2014 have resulted in humanitarian relief operations at national and international levels. Decision making by relevant actors in all aspects of humanitarian response impacts on the well-being of People of Concern (PoC). Site management decisions take into account a wide range of criteria in diverse sectors, including protection of PoC. Criteria address basic relief response, such as water, sanitation and nutrition, social characteristics, as well as protection of PoC. In particular, site management decisions may affect relationships between PoC, leading to tensions or peaceful coexistence between PoC. In this paper a decision making process is proposed for the forecasting of emotions in PoC sites in order to assist site management decisions, particularly in the context of avoidance of tensions. The method uses Fuzzy Cognitive Maps (FCM) to forecast emotions based on input data obtained from site management reports. Historical data from site management reports in Greece have been used for deriving the indicators used in the input layer of the designed FCM. The proposed method will be applied to forecast potential tensions in PoC sites in Greece.

Keywords: Artificial emotion forecasting · Fuzzy cognitive maps · Refugee site management · Decision support method · Refugee crisis

1 Introduction

After the massive sea arrivals in the Mediterranean in the period 2015–2016, the numbers of refugees and migrants (PoC) have dropped significantly. Thus, in 2017 a total of 172,301 PoC arrived by sea, whereas in 2018, as of November, 113,539 PoC arrived by both sea and land in the Mediterranean [1]. However, at a country level, the total arrivals in Greece in the time period January to October 2018 reached 41,252, an increase by 30% with respect to the same period in 2017. Moreover, in October 4,100 sea arrivals were registered, half of which at the island of Samos, where the number of PoC residing at the Reception and Identification Centre (RIC) of the island outreached

© Springer Nature Switzerland AG 2019
P. S. A. Freitas et al. (Eds.): EmC-ICDSST 2019, LNBIP 348, pp. 3–14, 2019.
https://doi.org/10.1007/978-3-030-18819-1_1

its capacity by six times. Currently 67,100 PoC reside in Greece, whereas 17,900 reside in the islands [1]. Protection warnings have been issued by humanitarian organisations including United Nations High Commissioner for Refugees (UNHCR) due to inadequate housing and the overcrowding conditions in the PoC sites at the islands as well as for many sites in mainland.

Humanitarian response and relief in all sectors, including settlement and shelter, requires coordination and collaboration between involved stakeholders including refugees and migrants and the host communities, local governments, municipalities, UNHCR, International Organisation for Migration (IOM) and other UN organisations, national and international Non-Governmental Organisations (NGOs), private and civic sectors and donors. In Greece, sites include RICs, official sites such as camps, as well as accommodation in buildings, hotels and apartments.

Settlement planning standards in response to crises and disasters have been issued by the SPHERE project [2], as well as UNHCR [3]. Accordingly, a set of planning standards ensure the social, economic and environmental sustainability of the operations that aims to ensure protection, health and well-being of PoC communities, in harmonisation with the host communities. Failure to meet the minimum planning standards could expose the refugee communities to security and protection risks, health risks, as well as create tensions [2, 3]. The minimum planning standards for shelter and settlement include covered living area, camp settlement size, fire safety, site gradient and topography [2, 3]. Decisions should adopt a "bottom-up" approach, i.e. focus on the needs of the needs of PoC at the individual family level, and consider water and sanitation, nutrition, health and education, waste management, communal services, among others. Moreover, in the context of security and protection, site planning and site management decisions should take into account the social structures, ethnic and cultural affinities between PoC such as ethnic groups and tribes, as well as age, gender and diversity, and their preferences such as in settlement layout and nutrition. Additional considerations should involve the relations and links of PoC with host communities. Therefore, site planning and management should encourage affinities and reduce or mitigate potential tensions and friction between PoC, as well as between PoC and host communities.

Research in humanitarian operations has grown significantly since the Asian Tsunami in 2004. However, research based on quantitative methods did not follow with the same rate [4] until recently. Drakaki et al. [5, 6] pointed out that the complexity of refugee site planning and siting can be addressed with an intelligent multi-agent system (MAS) that uses multi-criteria decision making (MCDM) methods [5, 6]. MCDM methods employed by agents in the MAS included the fuzzy analytic hierarchy process (FAHP) and hierarchical fuzzy axiomatic design (HFAD) with risk factors (RFAD). Land, location and supportive factors criteria as well as risk factors were considered by the decision support method. The MAS global goal was refugee site location identification based on evaluation and ranking of alternatives (sites). The method was applied to evaluate refugee sites in Greece. The authors made a comparative analysis of the developed MAS with a MAS that employed different MCDM methods, i.e. technique for order preference by similarity to ideal solution (TOPSIS) and fuzzy axiomatic design (FAD) [7]. Çetinkaya et al. [8] addressed refugee camp siting with a method that combined geographic information system (GIS) with FAHP. The authors used GIS to obtain potential refugee camps based on previously identified geographic, social, infrastructural and risk related criteria. FAHP was then used to determine weight values of GIS layers. Thus, alternative locations were determined. Subsequently, TOPSIS was used to obtain the ranking of alternative sites.

Human behavior is greatly influenced by emotions, therefore, Plutchik [9] argued that emotion is a social regulation process, a negative feedback homeostatic process, where behavior regulates progress to equilibrium. Models of emotion have conceptualized emotions in a way analogous to a color wheel [9, 10, 13], i.e. "similar" emotions are placed close to each other, whereas "complementary" emotions are placed at 180° to each other. Figure 1 on the left side, shows Russell's emotion model [10], adopted by Guojiang [11]. Salmeron [12] claimed that artificial emotion should be embedded in the reasoning module of intelligent systems that emulate or anticipate human behavior. The authors developed a method to forecast artificial emotions based on Thayer's model of emotions [13] and FCM. Kowalczuk and Czubenko [14] presented a review of computational models of emotion. The authors argued that after the system of needs, the mechanism of emotions constitutes a second human motivation system. Horn [15] studied the emotional and psychological well-being of refugees in Kakuma refugee camp. The author claimed that refugee emotions depended on current as well as past stressors. She suggested that response programmes addressing practical needs, especially safety and material needs have impact on psychosocial well-being. Therefore, she suggested that anti-social behavior in refugee camps could be related to refugee emotional problems. Drakaki et al. [5, 6] included risk factors in their decision support method for refugee siting. The considered risk factors have been included in Table 1, whereas they account for potential tensions in refugee sites as a result of overcrowding, as well as the type of PoC.

Therefore, in this paper, a method is presented to forecast emotions of PoC in sites that could lead to tensions, due to site conditions and site management response, in order to assist site management decisions on protection. An FCM is developed to forecast the emotions, whereas historical data obtained from site management reports are used to identify concepts of the FCM. The FCM concepts have been decided from a list of site management indicators which could lead to tensions if they are not met, based on UNHCR site management reports in Greece [1]. The proposed method is based on Russell's.emotion model [10] as adopted by Guojiang [11]. The method will be applied to forecast emotions in PoC sites in Greece.

In the following, background information on FCMs is presented next. The proposed method is described in the following section. Finally, conclusions are given which include future research.

2 The Emotion Model

Russell [10] claimed that emotions can be considered as dimensions which are interrelated and can be represented in a spatial map, where concepts (emotions) fall into a circle [22]. An example is shown in Fig. 1, adopted from [11]. The horizontal axis represents valence (or hedonic value) and can be considered as representation of pleasure/positive (at 0°) versus displeasure/negative (at 180°). The vertical axis represents arousal and can be considered as representation of "arousal" (at 90°) versus representation of "sleepiness" (at 270°).

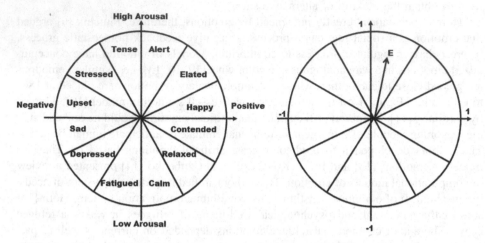

Fig. 1. The circumplex emotion model adopted from Guojiang [11] based on Russell's emotion model [10].

Arousal (E1) (x-axis) and valence (E2) (y-axis) can be represented by a vector, E [11]. The values of E fall in a unit circle, shown in Fig. 1, on the right side. E is calculated as

$$E = [E_1, E_2] = (\psi, \theta_E) \tag{1}$$

$$\psi = \sqrt{\frac{E_1^2 + E_2^2}{2}} \tag{2}$$

$$\theta_E(rad) = \begin{cases} arctan\frac{E_2}{E_1}, E_1 > 0 \\ arctan\frac{E_2}{E_1} + \pi, E_1 < 0 \\ \frac{\pi}{2}, \quad E_1 = 0 \end{cases} \tag{3}$$

3 Fuzzy Cognitive Maps

FCMS have been developed by Kosko [16] as an extension to cognitive maps. They are based on both fuzzy logic and neural networks. Complex dynamic systems characterised by abstraction and fuzzy reasoning can be modeled by FCMs, whereas both static and dynamic analysis of the modeled systems can be performed. A system is modeled as a directed weighted graph of interconnected nodes (concepts) with the connections between nodes showing the cause-effect relationships between the concepts. The direction of the connection between nodes shows the direction of causality whereas the value of the weight of the connection shows the amount of influence of the interconnection between nodes. The sign of the weight indicates whether the influence between concepts is positive or negative. Human knowledge by an expert or by a group of experts or historical data can be used to construct the configuration of the map. The development phase of an FCM includes three main steps, namely (i) the identification of important concepts, (ii) identification of causal relationships between the concepts and (iii) estimation of the strength of causal interconnections [17]. Domain experts decide on the values of the causal relationships (influences). They use fuzzy linguistic terms which are then mapped to numerical values in the range $[-1, 1]$.

FCMs have been used for decision support in diverse domains such as in [18–21]. Two main methods can be applied for decision support, namely static analysis which allows exploration and determination of the causal effects between concepts and dynamic analysis that allows the evolution over time of the modeled system until the FCM either stabilises to a fixed state, or shows a cyclic behavior or exhibits an unstable behavior.

Consider an FCM which consists of N concepts, C_i, where $i = 1,..., N$. Each concept, i, has a value in either $[0, 1]$ or $[-1, 1]$. The weights on edges, w_{ij}, have values in the interval $[-1, 1]$, where w_{ij} shows the influence of concept (cause node) i on concept (effect node) j. Positive influence means that an increase in C_i will cause an increase in C_j, a negative influence shows that an increase of C_i will cause a decrease in C_j, whereas $w_{ij} = 0$ indicates that there is no relation between concepts (nodes) i and j.

The weight matrix is formed as

$$\mathbf{W} = \begin{pmatrix} w_{11} & \cdots & w_{1N} \\ \vdots & \ddots & \vdots \\ w_{N1} & \cdots & w_{NN} \end{pmatrix} \tag{4}$$

Each concept value, C_i, is updated through an iteration procedure [18]. In iteration $k + 1$, C_i is updated according to

$$C_i^{(k+1)} = f\left(C_i^{(k)} + \sum_{j=1}^{N} w_{ji} C_j^{(k)} \right) \tag{5}$$

The function f is a threshold function. When the sigmoid function, $f(x) = \frac{1}{1+e^{-\lambda x}}$, is used concept values are in the interval $[0, 1]$, whereas when the hyperbolic tangent, $f(x) = \tanh(x)$, is used values are in the interval $[-1, 1]$. Iteration continues until a convergence is achieved.

4   The Proposed Methodology

The purpose of the decision support method is to assist site management decisions in PoC sites in order to avoid tensions among PoC or between different ethnic groups. The designed FCM follows a structure introduced by Salmeron [12], i.e. it consists of an input layer, a "hidden" layer and an output layer. The computational model of emotion introduced by Guojiang [11] based on Russel's emotion model [10] is used to map output nodes to emotions. The developed FCM is adopted for site management support in the Mediterranean. The development procedure consists of the following steps:

Step 1: Historical data from site management reports where tensions have been reported are used.
Step 2: Indicators to site management operations from the considered historical data which could contribute to tensions are identified.
Step 3: Concepts for the input and hidden layers are identified. Their causal relations, expressed with the arcs connecting nodes, as well as the sign of the corresponding weights are determined.
Step 4: The weight matrix is calculated.
Step 5: Values of output nodes, namely arousal and valence, are mapped to emotions according to the procedure introduced by Guojiang [11].

Site Management Operations

Site planning standards in response to disasters and crises have been issued by UNHCR in accordance with the SPHERE project [2]. Drakaki et al. [5, 6] presented site selection and planning criteria from UNHCR, the SPHERE project and academic literature. Site selection and planning criteria and corresponding standards cover strategic settlement planning, environment, construction, the needs of vulnerable groups, as well as risks. A range of standards ensure that PoC will not be exposed to risks related to life, health, security and protection, including tensions among PoC groups as well as between PoC and host communities. Accordingly, site management involves response to a range of factors, indicatively listed in Table 1, as adopted from UNHCR [1]. Human factors [6] that could attribute to the tensions, especially between ethnic groups, such as numbers, type of PoC, as well as population density are listed in Table 1. Risks associated with the standards are included in Table 1. The listed site management factors relate to site management response in the Mediterranean.

For the purposes of this study indicators to potential tensions from the site management response have been drawn from historical data from UNHCR in Greece, in the period between June 2017 to September 2018. They have been identified from site management reports from sites operating in Greece [1]. Specifically, site reports in which tensions either often or rarely have been recorded were identified. Indicators in these reports which were below the minimum standards were registered. Human factors listed in Table 1 could contribute to tensions. Table 2 shows indicators as well as human factor related indicators which were identified as potential contributors to the creation of tensions. These indicators form the input data to the input layer of the designed FCM. All indicators in Table 2 are checked in both columns. Individual indicators were found to be below the minimum standards both in sites where tensions appear rarely and in sites where tensions appear often. However, in the future, data could show different results.

Table 1. Site management response operations and risks associated with the minimum standards.

Site overview	Population and capacity
• Location/type of settlement • Fire safety • Check in and check out mechanism in place at entrance • Provision of electricity • Environmental hazards	• Estimated number of PoC hosted in site • Estimated number of potential new PoC able to reside in the vacant accomodations • Nationality breakdown (%) • Age & gender breakdown (%) • Average household size
Shelter	**Wash**
• Number of tents/number of PoC living in tents • Number of prefabricated housing units/number of accommodated PoC/cooling & heating/water & sanitation (WASH) facilities • Number of buildings/number of PoC living in buildings/provision of heating/cooling	• Communal WASH facilities • Provision of hot water • Number of functional showers • Showers available in separated areas for women • Wash facilities designed for people with disabilities • Access to potable water • Sewage • Garbage disposal
Food	**Health**
• Is food distributed?/frequency of distribution • Specific nutritional needs considered	• Primary health care available/MHPSS available • Referral system in place • Distance to the nearest public health facility/hospital
Protection	• SOPs in place
• UASC safe zone • Referral mechanisms • Restoring family links service • Legal assistance • Tensions between communities	• Private rooms for councelling • Illumination • Refugee community structure
Non-food items (NFI)	**Transport**
CASH	• Public transportation
Education	• Transport to regional asylum offices, embassies/schools provided
Communication with communities (CwC)	• Estimated distance to major cities
Risks	
Location-security, environmental risks	(Risks related to natural hazards, such as earthquakes, high winds, fire risks, flooding, landslide)
Location-security, health risks	(Locations that present health risks should be avoided)
Security and protection, high population density	(Increased exposure to health risks, tension, and protection threats to vulnerable groups)
Security and protection, the type of PoC	(Exposure of PoC to protection threats, tensions between ethnic groups)

Table 2. Site management indicators which could contribute to tensions in PoC, based on UNHCR reports in Greece.

Site management indicators	Tensions-rarely	Tensions-often
Security: check in and check out mechanism in place at entrance	X	X
Population and capacity: estimated number of potential new PoC able to reside in the vacant accomodations (capacity for new arrivals)	X	X
Population and capacity: nationality breakdown (%)	X	X
Population and capacity: age and gender breakdown (%)	X	X
Basic needs: shelter cooling and heating	X	X
Basic needs: WASH	X	X

The hidden layer nodes represent the concepts nervousness and well-being. Output concepts are arousal and valence, using the two-dimensional computational emotion model, shown in Fig. 1, introduced by Guojiang [11] adopted from Russell's model of emotion [10].

Based on Table 2, the input FCM layer consists of the nodes (concepts): security: check-in and check-out mechanisms, basic needs: (shelter) cooling and heating/WASH, population and capacity: capacity for new arrivals, population and capacity (nationality breakdown), population and capacity: age & gender breakdown. Security, basic needs and population and capacity refer to site management indicators. Capacity for new arrivals is an indicator for overcrowding. Age and gender breakdown is considered in terms of the male percentage. Thus a value of 1.0 denotes 100% male population. Nationality breakdown represents the relative percentage of the three top nationalities hosted in the site. In Greece, the top three nationalities from January to September 2018

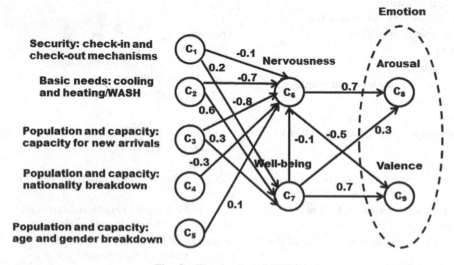

Fig. 2. The developed FCM.

are Syrian Arab Republic, Iraq and Afganistan [1]. A value of 1.0 represents 100% Syrian PoC. The hidden FCM layer consists of the nodes (concepts): nervousness and well-being. The output FCM layer consists of the nodes (concepts): arousal and valence. The developed FCM is shown in Fig. 2. Causal relationships between nodes are expressed as fuzzy numbers via the weights, which take values in the range [−1, 1] [17]. Linguistic terms were used to describe the causal relationships which were subsequently mapped to numerical values.

According to this process "no causal effect" between nodes is assigned to 0, full positive causality relationship is assigned to 1.0, full negative causality relationship is assigned to −1.0. The calculated weight matrix is shown in Table 3.

Table 3. The weight matrix.

Concepts	C_1	C_2	C_3	C_4	C_5	C_6	C_7	C_8	C_9
C_1 Security: check-in and check-out mechanisms	0	0	0	0	0	−0.1	0.2	0	0
C_2 Basic needs: cooling and heating/WASH)	0	0	0	0	0	−0.7	0.6	0	0
C_3 Population and capacity: capacity for new arrivals	0	0	0	0	0	−0.8	0.3	0	0
C_4 Population and capacity: nationality breakdown	0	0	0	0	0	−0.3	0	0	0
C_5 Population and capacity: age and gender breakdown	0	0	0	0	0	0.1	0	0	0
C_6 Nervousness	0	0	0	0	0	0	0	0.7	−0.5
C_7 Well-being	0	0	0	0	0	−0.1	0	0.3	0.7
C_8 Arousal	0	0	0	0	0	0	0	0	0
C_9 Valence	0	0	0	0	0	0	0	0	0

Historical data from a range of PoC sites in Greece have been used to produce the weight matrix, from June 2017 to September 2018 [1]. The sites used to calculate the weight matrix cover sites where tensions have been reported to exist either rarely or often. The data were collected from UNHCR reports for the Mediterranean situation [1], in the document sub-categories of factsheets as well as site profiles, for Greece. Site profiles indicate whether tensions exist in PoC sites, as well as the site management indicators in all sectors. PoC are exposed to increased tensions when indicators do not meet the minimum standards [3, 6]. Furthermore, monthly country factsheets provide data including statements about tensions, such as statements about overcrowding (linked to C_3) and unhygienic conditions or lack of heating/cooling (linked to C_2) leading to tensions. Additionally, the higher frequency of appearance of the above factors in statements linking them to tensions has been taken into account when assigning higher negative causality relationship values between C_2 and C_3 nodes and nervousness (C_6) in Table 3, with respect to the remaining input nodes (C_1, C_4 and C_5). The ethnic origin of different PoC groups (node C_4) should be considered as a factor that could contribute to tensions [6]. Furthermore, a high percentage of male population in node C_5 indicates lack of gender balance as well as separated families, and could indicate tensions. Nodes C_4 and C_5 represent the risk factor associated with type of PoC, which is listed in Table 1 as a source of potential tensions between PoC. Simulation scenarios will be explored to map emotions generated in PoC sites in Greece. The results will be tested based on real data from PoC sites in Greece in the same time period. Therefore, the weight matrix values in Table 3 will be adjusted and fine-tuned accordingly. The designed FCM is in line with Giardino et al. [23]. The authors identified two centers in the brain stimulating opposing emotional states. Accordingly, they argued that one center stimulates search for reward, whereas the other one stimulates fleeing from danger.

5   Conclusions

Human behaviour is highly influenced by emotions, therefore forecasting emotions could contribute to research in diverse areas including cyberphysical systems and human migration. Based on historical data from UNHCR site management reports, in the context of the PoC site management response, a range of factors could contribute to the emergence of tensions. In this paper a decision making process is proposed for the forecasting of emotions in PoC sites in order to assist site management decisions in the context of avoiding tensions in sites. Site management decisions affect the security and protection, health and well-being of refugees and migrants (PoC). The method uses FCM to forecast emotions. Historical data obtained from site management reports in Greece were used to identify the concepts of the input and hidden layer of the FCM. The developed FCM follows the structure of the FCM proposed in [12], whereas Russell's emotion model [10] as adopted by Guojiang [11] has been used for the identification of the (output) emotion. The method will be applied to forecast emotions in PoC sites in Greece that could lead to tensions in PoC sites. Future research will explore Particle Swarm Optimisation (PSO) to train FCM in order to avoid a subjective weight matrix calculation as well as the interrelationships between different indicators used in the input layer of the FCM. Furthermore, the method could be used to assist in human migration modeling, as well as in the identification of critical factors contributing to tensions between PoC and the host communities.

References

1. UNHCR data portal (2017). https://data2.unhcr.org. Accessed 17 Nov 2017
2. SPHERE: Sphere Project, Sphere Handbook: Humanitarian Charter and Minimum Standards in Disaster Response (2011). http://www.ifrc.org/docs/idrl/I1027EN.pdf. Accessed 20 Nov 2017
3. UNHCR emergency handbook: UNHCR Handbook for Emergencies. https://www.unicef.org/emerg/files/UNHCR_handbook.pdf. Accessed 20 Nov 2017
4. Gutjahr, W.J., Nolz, P.C.: Multicriteria optimization in humanitarian aid. Eur. J. Oper. Res. **252**, 351–366 (2016)
5. Drakaki, M., Gören, H.G., Tzionas, P.: An intelligent multi-agent system using fuzzy analytic hierarchy process and axiomatic design as a decision support method for refugee settlement siting. In: Dargam, F., Delias, P., Linden, I., Mareschal, B. (eds.) ICDSST 2018. LNBIP, vol. 313, pp. 15–25. Springer, Cham (2018). https://doi.org/10.1007/978-3-319-90315-6_2
6. Drakaki, M., Gören, H.G., Tzionas, P.: An intelligent multi-agent based decision support system for refugee settlement siting. Int. J. Disaster Risk Reduct. **31**, 576–588 (2018)
7. Drakaki, M., Gören, H.G., Tzionas, P.: Comparison of fuzzy multi criteria decision making approaches in an intelligent multi-agent system for refugee siting. In: Świątek, J., Borzemski, L., Wilimowska, Z. (eds.) ISAT 2018. AISC, vol. 853, pp. 361–370. Springer, Cham (2019). https://doi.org/10.1007/978-3-319-99996-8_33
8. Çetinkaya, C., Özceylan, E., Erbaş, M., Kabak, M.: GIS-based fuzzy MCDA approach for siting refugee camp: a case study for southeastern Turkey. Int. J. Disaster Risk Reduct. **18**, 218–231 (2016)
9. Plutchik, R.: A general psychoevolutionary theory of emotion in emotion: theory, research, and experience. In: Plutchik, R., Kellerman, H. (eds.), vol. 1. Academic Press (1980)
10. Mayne, T.J., Ramsey, J.: Emotions: Current Issues and Future Directions. The Guilford Press, New York City (2001)
11. Guojiang, W.; Xiaoxiao, W., Kechang, F.: Behavior decision model of intelligent agent based on artificial emotion. In: Advanced Computer Control (ICACC) (2010)
12. Salmeron, J.L.: Fuzzy cognitive maps for artificial emotions forecasting. Appl. Soft Comput. **12**, 3704–3710 (2012)
13. Thayer, R.E.: The Biopsychology of Mood and Arousal. Oxford University Press, Oxford (1989)
14. Kowalczuk, Z., Czubenko, M.: Computational approaches to modeling artificial emotion – an overview of the proposed solutions, frontiers in robotics and AI (2016). https://doi.org/10.3389/frobt.2016.00021
15. Horn, R.: A study of the emotional and psychological well-being of refugees in Kakuma refugee camp, Kenya. Int. J. Migr. Health Soc. Care **5**, 20–32 (2010)
16. Kosko, B.: Fuzzy cognitive maps. Int. J. Man Mach. Stud. **24**, 65–75 (1986)
17. Khan, M.S., Quaddus, M.: Group decision support using fuzzy cognitive maps for causal reasoning. Group Decis. Negot. **13**, 463–480 (2004)
18. Papageorgiou, E.I., Salmeron, J.L.: A review of fuzzy cognitive maps research, during the last decade. IEEE Trans. Fuzzy Syst. **21**, 66–79 (2013)
19. Xiao, Z., Chen, W., Li, L.: An integrated FCM and fuzzy soft set for supplier selection problem based on risk evaluation. Appl. Math. Model. **36**, 1444–1454 (2012)
20. Andreou, A.S., Mateou, N.H., Zombanakis, G.A.: Soft computing for crisis management and political decision making: the use of genetically evolved fuzzy cognitive maps. Soft. Comput. **9**, 194–210 (2005)

21. Papageorgiou, E.I., Hatwágnerb, M.F., Buruzsc, A., Kóczy, L.T.: A concept reduction approach for fuzzy cognitive map models in decision making and management. Neurocomputing **232**, 16–33 (2017)
22. Russell, J.A.: A circumplex model of affect. J. Pers. Soc. Psychol. **39**, 1161–1178 (1980)
23. Giardino, W.J., Eban-Rothschild, A., Christoffel, D.J., Li, S.-B., Malenka, R.C., de Lecea, L.: Parallel circuits from the bed nuclei of stria terminalis to the lateral hypothalamus drive opposing emotional states. Neuroscience **21**, 1084–1095 (2018)

The Use of a Decision Support System to Aid a Location Problem Regarding a Public Security Facility

Ana Paula Henriques de Gusmão[(⊠)], Rafaella Maria Aragão Pereira,
Maisa Mendonça Silva, and Bruno Ferreira da Costa Borba

Research Group in Information and Decision Systems (GPSID),
Universidade Federal de Pernambuco (UFPE),
Av. da Arquitetura, Cidade Universitária, Recife, PE 50740-550, Brazil
anapaulagusmao@cdsid.com, rafaellamaragao@gmail.com,
maisa@cdsid.org.br, brunoborba50@hotmail.com

Abstract. This paper aims to support decisions related to determining efficient spatial distributions of police units, given that the location of these units is a strategic matter regarding the costs of operations and police response times to occurrences. On conducting a literature review of the facility location problem in the public security area, it was noticed the mathematical model Maximal Covering Location Problem (MCLP) could be applied to a case study concerning military police units in Recife, a large city in Northeast Brazil. For the case study, MCLP, along with a Decision Support System (DSS), was applied to make an analysis of the potential location of military police facilities in Recife. This evaluation enabled the results obtained from an ideal scenario, where the bases were positioned at optimal points, to be compared with the performance of these facilities in their current location. For this purpose, CVP (an indicator for Violent Crimes against property georeferenced data from 2017 and the location of the military police units of Recife were used.

Keywords: Public security · Facility location problem · Decision Support System

1 Introduction

According to the 2017 Public Safety Yearbook [1], R\$ 2,314,708,998.81 was spent on the area of public security in Pernambuco in 2016, which represents 3.44% of the expenses of all public security units in Brazil. However, the number of victims of CVLI (an indicator for Violent Lethal and Intentional Crimes) in Pernambuco increased from 3,890 to 4,479 in 2016. This situation is more alarming when one considers that this number soared to 5,426 victims of CVLI in Pernambuco in 2017, a total that had never before been registered in the State. Crimes of this kind first started to be recorded in the Information System about Mortality of DataSUS in 1979. As to the number of CVP (an indicator for Violent Crimes against property) cases, there has also been a significant increase in recent years, namely, there were 84,945 cases in 2015, 114,802 in 2016 and 119,809 in 2017.

© Springer Nature Switzerland AG 2019
P. S. A. Freitas et al. (Eds.): EmC-ICDSST 2019, LNBIP 348, pp. 15–27, 2019.
https://doi.org/10.1007/978-3-030-18819-1_2

The SDS-PE (Secretariat of Social Development of Pernambuco) integrates the actions of the government of Pernambuco with a view to preserving public order and the safety of people and assets within the state and it is responsible for defining where police units are located. Its basic structure includes the Executive Department for Social Defense, the Executive Department for Integrated Management, the civil police, the military police (MP), the MP fire brigade, and management units and the superintendency which are administratively subordinated to the SDS. In the case of the superintendency, its work is technically linked to the Departments of Planning, Finance and Management and State Reform. The ostensible policing and the preservation of public order often falls to the MP, which uses the structure of its units to go to and attend emergencies.

Thus, with the aim of developing effective public security strategies, there is a need to better allocate the resources available for the policies for these in Pernambuco. In fact there are other initiatives to improve services and make better use of resources invested in public security, for example [2]. The idea of this paper is to present a study carried out to evaluate possible scenarios for the location of MP units, considering the demand for police services, distributed in the city of Recife, the state capital of Pernambuco, and the average time taken to attend them, thereby seeking to minimize the resources spent on providing these services.

To illustrate the applicability of the proposed DSS, 1,840 georeferenced records of police occurrences in Recife were examined which are held on the collaborative platform *Onde Fui Roubado (Where I was robbed)*. This platform maps occurrences of robberies, thefts and other types of crime in Brazilian cities. Thus, the focus was on where the MP installations in Recife are located. The Maximal Covering Location Problem (MCLP) approach was applied and some scenarios were constructed from different combinations of values for the parameters: service distance and the number of occurrences covered. Therefore, it was possible to compare the different scenarios with the current location of MP units and to suggest some improvements. Thus, the contribution of this paper is to support important decision-making in public security management: the allocation of MP units to the most appropriate locations.

2 Literature Review

2.1 Decision Support in Public Security

It has been proven that studying historical data enables places to be identified where crimes tends to agglomerate. Therefore, making use of historical information is fundamental for decreasing crime, as crime has greater chances of happening when there is evidence of why and where criminal activity is likely to take place and yet no local security measures are in place [3, 4]. Consequently, in the last decade, predictive policing measures have been developed with the aim of providing an analysis of the evolution of crime in a territory. More recently, both academics and practitioners, such as the RAND corporation and the NIJ (National Institute of Justice of the United States), have recognized the need to take a step forward and develop a DSS to provide help to decision makers (DMs) in law enforcement agencies [5].

A DSS was proposed in [6] so as to implement a new paradigm of predictive police patrolling for the efficient distribution of police officers in a territory under the jurisdiction of a police department, with the aim of reducing the likelihood of criminal acts. A DSS was also developed in [7] so as to plan police rosters that allowed a better sizing of the work force in the city of San Francisco and allowed a 25% increase in the number of workers, thereby decreasing in 20% the response time to the emergency service by 20% and reducing expenditure on the public security area by US$5.2 million per year.

On the other hand, Curtin et al. [8] have worked on the p-median problem to improve service coverage by cars in order to reduce travel time. In this context, the location of MP units at strategic points is fundamental if smaller operational costs and quicker response times to attend to occurrences are to be achieved. For this purpose, operational research methods and techniques are important tools in solving this type of engineering and planning problem [9].

2.2   Facility Location Problem

The Facility Location Problem (or Plant Location Problem or Site Location Problem) has given rise to countless studies in the area of linear and continuous optimization in the last 40 years. It is important to note that the term Facility is used to address the location problem generically, which may include the location of a factory, warehouse, trade, hospital, nuclear power plant, etc. The researchers focused on finding models and algorithms for problems related either to the private sector (industrial plants, banks, retail facilities, etc.) or to the public sector (ambulances, clinics, police units, etc.).

The term Location Analysis refers to modeling, formulating and solving a class of problems that can be described as positioning operations in any given space. There is a distinction between location and layout problems. The operation in location analysis is small compared to the space in which it is located and interactions with other operations may or may not occur. On the other hand, in layout problems operations are much larger than the space they are in and interactions with other operations are more of a rule than an exception.

Location problems are characterized by four components: (1) clients, who are located at points or on routes, by presupposition, (2) operations, which will be positioned, (3) space between clients and operations, and (4) measurements, which indicate distances or times between clients and operations. Scientists typically distinguish location problems according to the space they are in, i.e. in real D-dimensional or network space, each of which can be further subdivided into continuous problems (any point is feasible for a new facility) or discrete (there is a candidate set of viable points for a new facility).

In Network Location Problems, operations are positioned on the network nodes themselves, and typically, the shortest routes in the network of arcs are measured, which connects all the relevant pairs of points. Most discrete location problems are network ones and involve zero-one variables, which results in integer programming and combinatorial optimization problems. It is noteworthy that there are numerous hybrid models that do not fit the above classifications.

Another factor that is relevant in location modeling is the DM's objective. Traditionally, the operations to be positioned would have a higher value in the objective function, the closer they were to their clients. This case, according to [10], falls into the category of pull objectives. However since the late 1970s, researchers have taken into account the location of undesirable operations, since one of the goals of clients is to be as far away as possible from these operations. The latter case is classified as push objectives. A third class of objectives, known as balancing objectives, would be the scope of equity, which attempts to locate facilities in such a way that the distances between customers and facilities are as similar as possible to one another. Alternatively, distances can be delimited by generally recognized distance patterns.

The most common location models try to locate a single operation, while more complex ones involve more than one operation and the number of operations can be predetermined or determined endogenously by using the elements of the model.

Finally, [11, 12] distinguished location problems of the private sector and the public sector. The private sector aims to optimize some monetary function associated with the location; in contrast, the public sector seeks to optimize the population's access to its services. Normally setting goals in public sector models is much more complicated than in the private sector, since the latter often concentrates on maximizes profits and minimizing costs while the former is often more concerned about trying to define and meet intangible objectives.

2.3   Covering Problems

Although there were already some articles on network-based facilities and warehouse locations in the 1950s, it was only in the 1960s that this kind of problem received more attention. One of the pioneers in this area was [13] who investigated the minimum weighted distance of operations in a network with n nodes of demand, a classic problem he called the p-median problem. He did not present a solution method for this problem, but proved the existence of at least one optimal solution, which has all the operations located at the nodes of the network.

In addition to median problems, there are four other major categories of network problems: center problems, hub location problems, hierarchical location problems, and covering problems.

In some circumstances, particularly when emergency operations are to be allocated to a space, neither the concept of p-median, which has objectives of cost minimization and profit maximization, nor p-center is satisfactory. Instead, considering that DMs want to cover all customers, the concept of covering problems is more appropriate.

Generally, in covering problems, a customer or demand point must have coverage by at least one facility within a given critical distance. This critical distance is predefined and called the distance or radius of coverage [14]. The customer can receive services from any facility, the distance of which to the client is equal to or less than the default value. Therefore, the concept of coverage is more related to a satisfactory method than the best possible one. Various situations involving how to determine the number and location of public schools, police stations, libraries, hospitals, public buildings, post offices, parks, military bases, radar facilities, bank branches, shopping centers and waste disposal facilities can be formulated as covering problems [15].

[16] was the first to introduce covering problems by establishing a model to minimize the number of police officers needed to cover nodes of a road network. However, the first mathematical model, which was classified as LSCP (Location Set Covering Problem), was presented by [17]. An LSCP requires each consumer to have coverage of some facility within the standard distance, which is very restrictive.

In a problem with several spatially-dispersed demand nodes, this requirement can produce solutions with a number of installations that are unrealistic from the budget point of view. Thus, the MCLP (Maximum Covering Location Problem), introduced by [18], emerged. It seeks the economically viable number of operations (as the number of operations is limited, this is an exogenous problem) so that the number of customers covered is maximized.

The MCLP has the following basic structure:

$$Maximize\ Z = \sum_{i \in I}^{n} a_i y_i \tag{1}$$

Subject to:

$$\sum_{j \in N_i} x_j \geq y_i \quad \text{for all } i \in I \tag{2}$$

$$\sum_{j \in J} x_j = P \tag{3}$$

$$x_j = (0, 1) \quad \text{for all } j \in J \tag{4}$$

$$y_i = (0, 1) \quad \text{for all } i \in I \tag{5}$$

Where:

I = the set of demand nodes;
J = the set of potential operational locations;
$x_j = 1$ if an operation is located at node j, and 0 otherwise;
$y_i = 1$ if a demand node i is covered by at least one potential operational location j, and 0 otherwise;
$N_i = \{j \in J | d_{ij} \leq S\}$;
d_{ij} = the shortest distance between node i and node j;
S = maximum established value of the distance between the demand node and the operation node (desired service distance);
a_i = population served at demand node i;
P = number of operations to be allocated.

N_i is the set of operations sites eligible to provide coverage for a demand node i. A demand node is covered when the location of the operation closest to the node is less than or equal to S. The objective is to maximize the number of demand nodes covered within the desired service distance. Constraints of type (2) allow y to be equal to 1 only when one or more operations are established at locations within the set N (i.e., at one or more operations that are allocated within the distance S relative to the demand node i). The number of operations allocated is limited by P in constraint (3), which is defined by the user. Constraints (4) and (5) indicate that only integer values can be part of the solution. The solution to this problem specifies not only the percentage of the population that can be covered, but also the location of P operations that will be used to achieve maximum coverage.

From the literature review on the facility location problem, it was noticed that using a method such as MCLP can be an excellent way to allocate and analyze the coverage of the MP units of Recife.

An application of the MCLP in the public security area was found in [8]. These authors argued that, historically, the geographical boundaries of the police were delimited based on the knowledge of an administrator or officer of the total area to be patrolled and on the police resources. In some cases, there was concern about natural boundaries, hotspots, or demographic censuses, but generally speaking, there was no quantitative method by which the police could divide an area, which makes the efficient distribution of resources a difficult task, taking into consideration the need to reduce the response time and to make cost savings.

This pervasive and persistent lack of formal procedures to develop the police patrol area can complicate higher-level police decision-making [7]. In this perspective, they proposed to use the MCLP by integrating the GIS (Geographic Information System) and the incident database with linear programming software to generate and display optimal solutions. In addition, they have formulated an innovative Backup Coverage Model, which considers that more than one location of operation can cover a demand node within the service distance.

In [8], the authors formulated the MCLP based on criminal data and geographic boundaries of Dallas police in Texas and called this model the PPAC (Police Patrol Area Covering). Demand nodes are known incidence sites or service calls. The potential operation locations are the possible locations of police patrol command centers. The S can be either the service distance or the response time. The a_i is the weight or priority of criminal incidents at an incidence site i. The P can be limited by police features, for example, the number of patrol cars available, which is an advantage considering that these features can change frequently and quickly.

The PPAC method assumes that an acceptable service distance (response time) represents an adequate level of citizen safety. This assumption is reasonable, since police response time can be decisive in evaluating police performance. Operation sites are called command centers of the police patrol, because although police officers often answer calls while on patrol within their patrol area, there is no way that their position can be known in advance. Thus, a central point becomes the best assumption of the location of these police officers.

In the case of Dallas, the priority of incidents was established by using the Signal Code, which communicates the priorities assigned to the different calls received by the police and to response procedures. These calls range from extremely serious incidents (murders, armed robberies, wounded officers, etc.) to less serious incidents such as vandalism and car accidents without injured persons. The distance S used was 2 miles (and 1 mile for one of the divisions) according to the researchers' observation of the distance between two subareas of the area to which the command center would be allocated. Unfortunately, there is no single, well-accepted value for acceptable distance or police response time [19]. After the PPAC resolution, one of the relevant results obtained was an improvement of 18% in the coverage of incidents within the service distance used when compared to the police configuration of Dallas.

Based on [8], due to the similarity of the problem addressed, a DDS was developed to support the location of MP units in Recife and it is presented in the next section.

3   The Proposed DSS

The architecture of the DSS, used in this paper, is illustrated in Fig. 1.

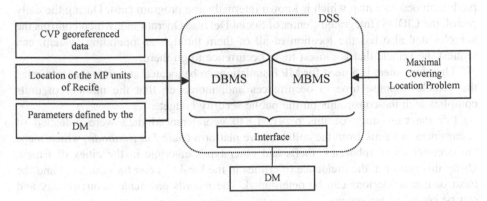

Fig. 1. The proposed DSS.

The main elements of the DSS are as follows:

- A database management system (DBMS): this stores and manipulates information regarding the public safety of different geographic areas of Recife and other data required to run the model. It comprises: the CVP georeferenced data of 2017 obtained from the platform *Onde Fui Roubado*; the location of the MP units of Recife and the parameters required by the model (service distance, response time, weights of the occurrences weight, for example), which are defined by the DM.
- A model-based management system (MBMS): this includes a MCLP model as described in Sect. 2.3 and the parameters required are defined by the DM (service distance, response time, weights of the occurrences, for example); and
- A user interface management component: this is the component that allows a user to interact with the system and to play an important role in a DSS. This element is under development.

The DSS was developed using the IBM ILOG CPLEX Optimization Studio software, to make an analysis of the potential locations of the MP installations of Recife. With the objective of improving the data storage and the interface with the DM, since the interaction is currently carried out via electronic spreadsheets, the DSS is being redeveloped using a more robust programming language in order to enable data to be dealt with better and the interaction to be more user-friendly.

4 Identifying the Potential Location of MP Installations

In this paper, a case study was conducted related to the location problem of MP units of Recife. According to Yin [20], a case study is characterized as an empirical study that investigates a current phenomenon in the context of real life. On the other hand, [21] emphasizes that the main benefits of a case study are the possibility of developing new theories and increasing understanding of real and contemporary events.

MP units are the places where MP officers perform their duties, and consequently make up the scope of the military police. On a daily basis, at each shift (dawn, morning, afternoon or night) vehicles leave these units to patrol an area that has already been predetermined on a map which is known internally as a program card. During the daily patrol, the CIODS (Integrated Center of Social Defense Operations), which controls the vehicles and also has the location of all of them through an operating system, can request the vehicle that is closest to an occurrence to go there.

Therefore, there is a need for MP installations to be located at points that minimize the vehicle response time to occurrences and, moreover, that the number of units complies with the constraints on the public security budget.

For the case study of this paper, 1,840 georeferenced data were collected of occurrences in Recife from the collaborative platform *Onde Fui Roubado*, which maps the occurrences of robberies, thefts and other types of crime in the cities of Brazil. Using this platform, the incidence of crimes in the localities can be visualized and the most dangerous regions can be determined. The records are made anonymously and can be consulted by anyone.

The occurrences collected fit the criminal CVP indicator, which covers all crimes classified as robbery, extortion by means of kidnapping and robbery with restriction of the victim's freedom, except for robbery followed by death which is accounted for in the CVLI indicator. According to the commander of the military police of Pernambuco, the CVP is in fact one of the relevant criteria for deciding to locate MP units. The other criteria are CVLI, population, territorial extension and presence of other units in the area.

The data cover the period between January 1, 2012 to July 14, 2017. However only the occurrences in 2017 were considered for the DSS application. The main reason is that *Onde Fui Roubado* was not a very well-known tool until recently. For the application of the MCLP, the 365 occurrences registered from January 1, 2017 to July 14, 2017 formed the set of demand nodes and part of the DBMS. As there were no data on potential operational sites, the centroids of the Recife neighborhoods that had the highest number of CVP cases were used.

From Table 1 and considering the Demographic Census, it was noticed that of the ten most populous neighborhoods, five (Boa Viagem, Várzea, Imbiribeira, Iputinga and Cordeiro) were considered in the viable points of the MCLP model, which is an important piece of information, since the size of the resident population is also a criterion for deciding where to locate MP units. As a budget constraint, of the ten potential operation sites no more than five points were selected. The choice of only five facilities occurred because of the initial interest in comparative analysis with the five MP units in Recife.

Table 1. Number of occurrences by neighborhood.

Neighborhoods	No. of CVP occurrences
Boa Viagem	63
Boa Vista	30
Várzea	26
Madalena	24
Santo Amaro	17
Santana	16
Imbiribeira	13
São José	12
Iputinga	11
Cordeiro	10

After selecting the demand nodes and potential operational locations, the distance between each pair of nodes was calculated. The Euclidian distances were calculated using the Haversine formula, an important equation used in navigation, which provides distances between two points of a sphere from their latitude and longitude. A conversion factor was used to provide distances in kilometers.

Regarding the desired distance parameter, it has already been mentioned that there is no well-accepted value for a service distance or a response time. It falls to the DM to define the parameters in line with the public safety goals of the locality. Thus, values of between two and five kilometers were tested. Finally, no priority was established among the 365 occurrences, i.e., each and every demand was considered of equal importance (weight). The MCLP model adapted to the reality of the public security management of the city of Recife formed the MBMS. Once all necessary inputs were at hand, a code for MCLP resolution was made, using the IBM ILOG CPLEX Optimization Studio software, and the results are presented in Table 2.

Table 2. Results of the objective function of the MCLP for the allocation of five MP units to the most appropriate locations.

Service distance (km)	No. of occurrences covered	% of occurrences covered
2.00	207	56.71%
2.50	256	70.14%
3.00	284	77.81%
3.50	315	86.30%
4.00	348	95.34%
4.50	355	97.26%
5.00	359	98.36%

From Table 2, it was seen that a service distance of 3, 3.5 and 4 km provides a reasonable coverage of the occurrences, namely, 77.81%, 86.3% and 95.34%, respectively. Therefore, it was decided to compare the distance traveled to the incidents by the vehicles that leave the facilities that have been allocated to optimal points according to Table 3 and the distance travelled by the vehicles that leave the existing facilities. For this purpose, georeferenced data were collected from the five existing units and added to the DBMS, as were the territorial responsibilities of these units.

Table 3. Operation sites of the MCLP for the allocation of five MP units.

Service distance (km)	No. of occurrences covered	Operational sites
2.00	207	Santo Amaro, Santana, Imbiribeira, São José and Cordeiro
2.50	256	Santo Amaro, Santana, Imbiribeira, São José and Cordeiro
3.00	284	Santo Amaro, Santana, Imbiribeira, São José and Iputinga
3.50	315	Boa Viagem, Santo Amaro, Imbiribeira, São José and Cordeiro
4.00	348	Boa Viagem, Santo Amaro, Imbiribeira, São José and Iputinga
4.50	355	Boa Viagem, Boa Vista, Imbiribeira, São José and Iputinga
5.00	359	Boa Vista, Várzea, Santana, Imbiribeira and Cordeiro

For the calculation of the total distance traveled, the shortest distance to the occurrence (d_{ij}) was used independently of the territorial responsibility of the unit in question. This actually occurs in practice, since the CIODS requests the vehicle that is the closest to the occurrence to go there. Table 4 presents the number of occurrences covered and the total distance covered, for three scenarios of service distance, using the suggestion of the DSS (based on MCLP) regarding the five operational sites.

Table 4. Number of occurrences covered and the total distance covered using the suggestion of the DSS regarding the five operational sites.

Service distance (km)	No. of occurrences covered	Total distance covered (km)
3.00	284	784.1529673
3.50	315	840.1488102
4.00	348	852.1651609

For comparison, Table 5 presents the performance of the same indicators (number of occurrences covered and the total distance covered) for the (apagar: to the) current location of the MP units.

Table 5. Number of occurrences covered and the total distance covered by the current location of the MP units.

Service distance (km)	No. of occurrences covered	Total distance covered (km)
3.00	256	920,5915959
3.50	285	920,5915959
4.00	306	920,5915959

According to Table 4, considering a service distance of 3 km, for the facilities allocated to optimal points the total distance covered was 784.15 km while for the current units (Table 5) the total distance covered was 920.59 km. There is also a difference in the number of occurrences covered for each facility which was obtained by the MCLP. Considering, for example, a service distance of 3.00 km, using the suggestions of the DSS, 284 occurrences are covered, as shown in Table 2, while the current units covered 256 occurrences for this distance. These differences demonstrate that the use of the MCLP can help the DM to allocate MP units to locations in a manner that reduces costs and makes attending to the occurrences more efficient.

A problem was perceived considering the distribution of territorial responsibility. An example is that all incidents in the neighborhood of Afogados, which should be served by the 12° BPM, are closer to the 19° BPM, located in Pina. This problem is a consequence of the lack of formal procedures related to efficient spatial distributions of MP units. In addition, the possibility of adding another MP unit was examined in order to increase the number of occurrences covered. Therefore, the constraint in the MCLP model was changed to allocating six units to the most appropriate locations. The percentage increase observed in Table 6 refers to the relative increase in the number of occurrences covered when six MP units are selected instead of five.

Table 6. Number of occurrences covered and the total distance covered using the suggestion of DSS for six MP units.

Service distance (km)	No. of occurrences covered	Percentage increase
2.00	215	3.86%
2.50	259	1.17%
3.00	291	2.46%
3.50	321	1.90%
4.00	348	0.00%
4.50	356	0.28%
5.00	359	0.00%

These results show that this percentage increase does not exceed 4% for any of the service distances tested. In addition, when the service distance is four or five kilometers there is no increase in the number of incidents covered. Thus, it is not feasible to allocate more MP units to other locations considering the data used. This is because the construction costs of a new MP unit are not likely to be compensated for by an increase in the coverage of occurrences.

5 Conclusions

The efficiently spatial distribution of police units, such as military police units, is a relevant point for improving the services provided by the public security area. Especially if it is taken into account that the successive increases in criminal indicators indicate the need for a faster response by police vehicles to the sites of occurrences.

Although every MP department has its criteria to define the allocation of its resources, there is a shortage of quantitative methods to analyze these spatial divisions. This paper demonstrates that this allocation of MP facilities to optimal points can be made by using an operational research model such as the MCLP.

The case study in Recife showed that by making an optimal allocation of the MP units, the distance that vehicles travel to incidents can be reduced. In addition, it has been found that reducing the number of facilities and redistributing resources may be a more viable solution than simply increasing the number of MP units.

The main contribution of this paper is to propose a DSS based on a mathematical model to support strategic and operational decision-making in the public security field. The model takes into consideration the budgetary constraints of this field and is flexible with regard to changing some parameters, such as weight and service distance desired, and to analyzing several scenarios.

To provide better analysis of the results, the use of CVLI georeferenced data has been developed and weights have been used to establish the priority of occurrences. To complete the DSS structure, the user interface is also is being improved.

Future lines of research can usefully further explore this same problem, by aggregating importance to the different occurrences and/or by using other approaches.

Acknowledgment. This research was partially supported by the Federal University of Pernambuco (UFPE) and the Brazilian Research Council (CNPq) by the CNPq Process: 422785/2016-4 - CHAMADA UNIVERSAL MCTI/CNPq N° 01/2016.

References

1. Brazilian Public Security Yearbook, ISSN 1983-7364 year 11 (2017). http://www.forumseguranca.org.br/wp-content/uploads/2017/12/ANUARIO_11_2017.pdf
2. Gusmão, A.P.H., Costa, A.P.C.S., Silva, M.M.: A decision support system to define public safety policies. In: 2018 International Conference on Decision Support System Technology, Heraklion, ICDSST Proceedings 2018, (2018)
3. Tita, G.E., Cohen, J., Engberg, J.: An ecological study of the location of gang "set space". Soc. Probl. **52**(2), 272–299 (2005)
4. Tita, G.E., Ridgeway, G.: The impact of gang formation on local patterns of crime. J. Res. Crime Delinquency **44**(2), 208–237 (2007)
5. Perry, W.L., McInnis, B., Price, C.C., Smith, S., Hollywood, J.S.: Predictive Policing: The Role of Crime Forecasting in Law Enforcement Operations. RAND Corporation, Santa Monica (2013) http://www.rand.org/pubs/research_reports/RR233
6. Camachos-Collados, M., Liberatore, F.: A decision support system for predictive police patrolling. Decis. Support Syst. **75**, 25–37 (2015)

7. Taylor, P.E., Huxley, S.J.: A break from tradition for the San Francisco police: patrol officer scheduling using and optimization-based decision support system. Interfaces **19**, 4–24 (1989)
8. Curtin, K.M., Hayslett-McCall, K., Qiu, F.: Determining optimal police patrol areas with maximal covering and backup covering location models. Netw. Spat. Econ. **10**, 125–145 (2007)
9. Morais Gurgel, A., Pires Ferreira, R.J., Aloise, D.J.: Proposta de modelos para a localização de bases policiais e roteirização de viaturas. In: ENEGEP (2010)
10. Eiselt, H.A., Laporte, G.: Objectives in location problems. In: Drezner, Z. (ed.) Facility Location: A Survey of Applications and Methods, pp. 151–180. Springer, Berlin (1995)
11. Revelle, C.S., Marks, D., Liebman, J.: Analysis of private and public sector location problems. Manag. Sci. **16**, 692–707 (1970)
12. Revelle, C.S., Eiselt, H.A.: Location analysis: a synthesis and survey. Eur. J. Oper. Res. **165**, 1–19 (2005)
13. Hakimi, S.L.: Optimal locations of switching centers and the absolute centers and medians of a graph. Oper. Res. **12**, 450–459 (1964)
14. Fallah, H., Naimisadigh, A., Aslanzadeh, M.: Covering problem. In: Zanjirani Farahani, R., Hekmatfar, M. (eds.) Facility Location: Concepts, Models, Algorithms and Case Studies. Physica Verlag, Heidelberg (2009). https://doi.org/10.1007/978-3-7908-2151-2_7
15. Francis, R.L., White, J.A.: Facility layout and location an analytical approach, 1st edn. Prentice-Hall, Englewood Cliffs (1974)
16. Hakimi, S.L.: Optimal distribution of switching centers in a communication network and some related graph theoretic problems. Oper. Res. **13**, 462–475 (1965)
17. Toregas, C., Swain, R., Revelle, C., Bergman, L.: The location of emergency services facilities. Oper. Res. **19**, 1363–1373 (1971)
18. Church, R.L., Revelle, C.: The maximal covering location problem. Papers of the Regional Science Association, vol. 32, pp. 101–118 (1974)
19. Hill, B.: Crime analyst via crimemap. Listserve (2006)
20. Yin, R.K.: Estudo de caso – planejamento e método, 2nd edn. Bookman, São Paulo (2001)
21. Souza, R.: Case research in operations management. In: EDN Doctoral Seminar on Research Methodology in Operations Management, Bruxelas (2005)

Exploiting the Knowledge of Dynamics, Correlations and Causalities in the Performance of Different Road Paths for Enhancing Urban Transport Management

Glykeria Myrovali[1(✉)], Theodoros Karakasidis[2],
Avraam Charakopoulos[2], Panagiotis Tzenos[1], Maria Morfoulaki[1],
and Georgia Aifadopoulou[1]

[1] Hellenic Institute of Transport/Centre for Research and Technology Hellas
(HIT/CERTH), 6th km Charilaou-Thermi Road, 57001 Thessaloniki, Greece
`{myrovali,ptzenos,marmor,gea}@certh.gr`
[2] Laboratory of Hydromechanics and Environmental Engineering,
Department of Civil Engineering, University of Thessaly, 38334 Volos, Greece
`thkarak@uth.gr, avracha@yahoo.gr`

Abstract. The great abundance of multi-sensor traffic data (traditional traffic data sources - loops, cameras and radars accompanied or even replaced by the most recent - Bluetooth detectors, GPS enabled floating car data) although offering the chance to exploit Big Data advantages in traffic planning, management and monitoring, has also opened the debate on data cleaning, fusion and interpretation techniques. The current paper concentrates on floating taxi data in the case of a Greek city, Thessaloniki city, and proposes the use of advanced spatiotemporal dynamics identification techniques among urban road paths for gaining a deep understanding of complex relations among them. The visualizations deriving from the advanced time series analysis proposed (hereinafter referred also as knowledge graphs) facilitate the understanding of the relations and the potential future reactions/outcomes of urban traffic management and calming interventions, enhances communication potentials (useful and consumable by any target group) and therefore add on the acceptability and effectiveness of decision making. The paper concludes in the proposal of an abstract Decision Support System to forecast, predict or potentially preempt any negative outcomes that could come from not looking directly to long datasets.

Keywords: Big traffic data · Cross correlation · Granger causality ·
Decision support systems · Mobility patterns · Travel time · Floating taxi data

1 Introduction

We probably live just at the beginnings of the Big Data Era; cloud computing, sensors located around us, Internet of Things with the connected "smart" - even wearable - devices, social networks, other crowd sourcing and crowd learning channels provide new opportunities with hidden ones to be also expected. From the other side, and in order to gain as many as possible from big data benefits, challenges arising from the

© Springer Nature Switzerland AG 2019
P. S. A. Freitas et al. (Eds.): EmC-ICDSST 2019, LNBIP 348, pp. 28–40, 2019.
https://doi.org/10.1007/978-3-030-18819-1_3

acquisition of huge data volumes cannot be ignored; more than ever, the need for advanced techniques for data quality assessment [1], fusion for improving inferences' accuracy [2] and visualization [3] comes to the fore.

Traffic sector could not and should not stay as an impartial observer under the massive scales changes triggered by the explosion of big traffic data; modern real time traffic data collection systems (image and video processors, Bluetooth detectors, radar, floating car data, crowdsourcing initiatives) have enabled the enrichment of traditionally collected transport data (field measurements, inductive or magnetic loops, road tube counters, piezo sensors) while offering an overall picture of the traffic state for entire road networks [4–6].

Understanding the need for transforming big (traffic) data into coherent knowledge able to be utilized by humans (decision makers and travellers in this case), the current paper concentrates on finding out underlying processes of traffic spatiotemporal dynamics inside an urban environment by making use of advanced timeseries analysis methods.

Floating car data (GPS enabled), thanks to the ability of simultaneously recording among others time, speed and space information for many paths (segments) of the road network – basic points of applying complex network theory for urban traffic planning – and to the low cost and high coverage performance, has been the recent point of reference on traffic data and information collection. Accepting the proof of concept for the efficiency of a traffic monitoring system based on GPS-enabled traffic data collection, the basic input of the current work comes from a GPS based database - the timeseries datasets used in the current work refer to travel times for crossing specific road paths [7]. Fusion techniques thorough review is beyond the scope of this paper, however, literature is rich of proposed approaches [8–13].

The scope of the paper is twofold; first, to examine traffic patterns and road paths reactions to certain events and second to determine the level of correlation among travel times in different road paths that are basic components of a greater city transport network. In this way, we aggregated descriptive statistics for a large period and we employed cross-correlation coefficient and Granger causality analysis.

The remainder of the text is organized as follows: Sect. 2 describes the case study, Sect. 3 presents the methodological approach along with the materials that have been used while Sect. 4 discusses the main research findings.

2 Study Area and Input Data; Thessaloniki as a Smart Traffic Laboratory

2.1 The Study Area

The study area for the implementation of the proposed methodological approach is the wider area of Thessaloniki, the second largest Greek city, with a population of 800.000 inhabitants (2011). Daily trips in Thessaloniki are highly dependent on private vehicles use (67%), 23% of the trips are conducted by Public Transport while significant is also taxis modal share (4%). Finally, the remaining modal share accounts to 4% by motorcycles and another 2% with non-motorized modes of transport [14].

Thanks to the robust cooperation among core key actors in the city (Quadruple Helix – local and regional authorities, research/facilitators: CERTH/HIT[1], industry: taxi association - ICT enablers – private companies, citizens), Thessaloniki is considered a smart transport city and a live laboratory; floating car data from approximately 1200 taxis, multiple traditional sensors, cameras, radars and 43 point-to –point Bluetooth detectors create a strong data layer for the city traffic state.

2.2   Input Traffic Data

For the current work, the travel time timeseries are composed by GPS-enabled data from the taxi sample of size 1200 collected on an ongoing basis.

Each of the 1200 taxi vehicles is equipped with an on-board smart device with geolocation capabilities, directly connected (4G-LTE) to the taxi dispatching centre and sending information on speed and GPS location every (around) 6 s. The FCD data as received by CERTH/HIT are further processed depending each time on the purpose of service/knowledge the Institute wants to create. For the current work, the procedure for deriving valid travel times from coordinates consists of the following steps;

   i. the validity of data asks for applying map matching process which links each GPS location to a road link (GIS) taking also into account the on-going trajectory (topological backtracking trajectory consistency within the network).
  ii. the spatial filtering refers to the spatial join methodology using GIS defined polygons for defining a first clustering of the locations in the road types.
 iii. additional filtering of data has been applied in order to exclude invalid input (applying thresholds to the orientation and speed values) as well as eliminating 0-speed values in the proximity of the 3 taxi stands in the area (in order to avoid data from stationary taxis influencing the analysis).
 iv. finally, data was aggregated to 15-min intervals (for avoiding interpolation need) [15].

2.3   The Road Network

The analysis presented in the current paper is made per road path and comparative results were reached. The road paths selected are defined so as to have also available data from the point-to-point detectors/Bluetooth (for fusion reasons which is however out of the scope of the current paper but serves the scopes of CCCN approach as depicted in Fig. 2), therefore, paths are defined where BT devices are available (beginning and end) – path based approach is adopted [16]. Furthermore, the maximum possible length of a path was considered so as to consist a realistic subpart of a total trip for many travellers; 2 km is the largest path [17]. After considering also the available datastamps from the taxi vehicles and based on the knowledge of the transport system

---

[1] The Hellenic Institute of Transport (HIT) is part of the Centre for Research and Technology Hellas (CERTH) which is a non-profit organization that directly reports to the General Secretariat for Research and Technology (GSRT), of the Greek Ministry of Culture, Education and Religious Affairs. http://www.imet.gr/index.php/en/.

operation for the city (modal split, OD data, taxi usage), the final paths selected for the analysis are the paths presented in Fig. 1.

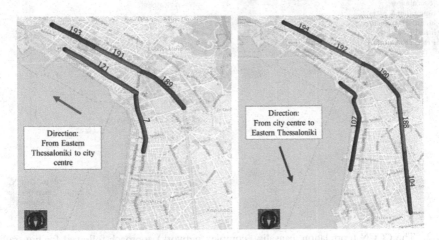

Fig. 1. Selected paths for the analysis of travel time timeseries

A first qualitative clustering of the paths shows that they are paths of similar characteristics (more than one lanes per directions, subparts of the main horizontal road axis serving road transportation needs of the city, with a length ranging from 1 to 2 kms where taxi share is relatively high).

3 Methodology

The deepening on multi-sensor traffic data understanding for the case of Thessaloniki, with generalization potentials, followed the stepwise approach depicted in Fig. 2.

The current paper concentrates on the 2nd step of the entitled **CCCN approach** (the green one) that refers to the estimation of cross correlations and Granger causality functions (and the respective knowledge graphs).

The previous step (grey) includes raw data pre-processing via simple statistical analysis (i.e. identification and treatment of outliers, zero values, recurring values, major differences among datasets from different sources) and an abstract map inference technique for taxi GPS coordinated allocation on the defined paths. Steps 3 and 4 that are out of the scope of the current paper, includes the transformation of travel data (here travel times) timeseries on complex network system [18, 19].

The last step of the CCCN approach is the development of a Decision Support System (DSS) that, through the identified relations among road paths and the dynamics of traffic, will work in two directions:

- Visualize spatiotemporal patterns
- Present the causal relation among traffic variables reaction at different road paths

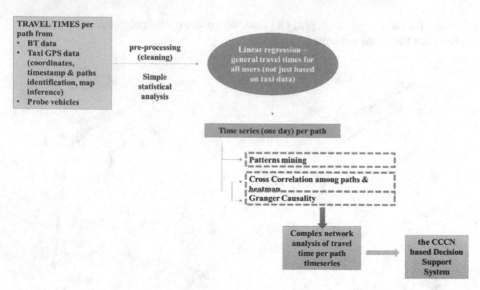

Fig. 2. The CCCN (correlation, causality, complex network) approach followed for big traffic data exploitation

Graphs deriving after each step of the CCCN approach feeds this last step of the DSS development.

3.1 Aggregated Statistics

In order to identify patterns and look into the interrelations between paths, a big dataset for a long period over a year was taken into account; taxi geolocalized timestamps for the 1st semester of 2017 was included in the initial data analysis. Average travel times at each day quarter (15-min analysis since the unavailable values were highly less when 10 min and 1 min analysis was attempted and therefore interpolation was necessary – therefore axis x depicts the 96 quarters of day) were calculated per daytype (weekdays, Saturday, Sunday) in all paths as well as timeseries per path were composed.

3.2 Cross Correlation Functions

The cross-correlation function (CCF) was applied to provide insight to the relationships among discrete paths timeseries by measuring the similarity between the first timeseries and the second one shifted as a function of time lag. We say that two time series (y_t and x_t), are correlated to one another when y_t is related past or future lags of x_t [20, 21].

3.3 Granger Causality

Since correlation does not always encompasses the concept of causality, the correlation analysis is completed with the Granger Causality analysis at the respective road paths pairs [22–25]. Supposing we have the same as above in correlation two time series

(y_t and x_t), we say that x_t Granger-(G)-causes y_t (or the opposite) if the past of x_t contains information that helps predict the future of y_t. For the purposes of the current work we employed the Conditional Granger Causality methods described in the 'Multivariate Granger Causality Toolbox' (MVGC) by Barnett and Seth [26]. For the calculation of the order of the model, we used the local minimum value according to either the Akaike or Bayesian information criteria.

4 Results and Discussion

Each subchapter of the current chapter presents the results of the respective methodological steps presented in the previous subchapters (i.e 3.1 aggregated statistics -> 4.1 initial step for patterns recognition).

4.1 Initial Steps for Patterns Recognition

Composing average daily travel time timeseries by the FCD data from the taxi for the specific road paths, the overall picture seems to verify the generic idea (intuition) and knowledge for city's traffic operation – different patterns in weekdays, Saturdays and Sundays – as results from the total population activity (with car trips dominating) (Fig. 3).

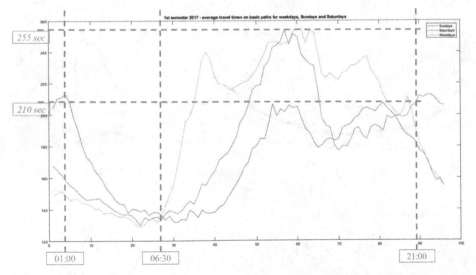

Fig. 3. Average travel time timeseries per day type (weekdays, Saturdays, Sundays) ave, 1st semester 2017

In Fig. 3, on the x-axis the time-increments/date is plotted (from 0 to 96 representing the respective quarters of day) and on the y-axis the corresponding value measured (travel time to cross the respective path/paths in seconds) of the presented timeseries is presented.

Taxi data for such a large period (central limit) verify also the peak periods during the weekdays (from 6:30 pm to 7:00 am) – more extended duration (bigger spread) inside which peaks appear - and the different profile for Saturdays and Sundays (leisure and shopping activities until the closure of the shops in the central area at early afternoon for Saturdays and night entertainment late night hours of Saturday and the early hours of Sunday).

From Fig. 4 we can see that peak hours for weekdays is 09:00 am, the period 1:30 pm–4:00 pm and around 7:00 pm (on average for all paths). For Saturdays the peak period is observed between 1:45 pm–4:30 pm while at this period we see travel times similar to these of the peak hours in weekdays (intense traffic). As for Sundays,

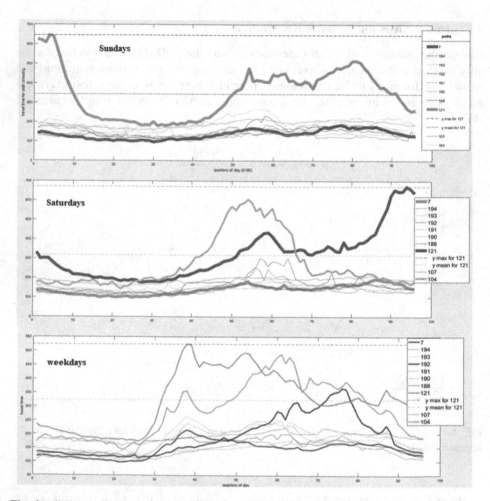

Fig. 4. (i) Timeseries per path, Sundays, 1st semester 2017, (ii) Timeseries per path, Saturdays, 1st semester 2017 and (iii) *Travel time timeseries (per path), weekdays, 1st semester 2017*

the travel times are notable lower, presenting peak over the early hours (1:00 am), the noon (1:00 pm–2:30 pm) and early evening (8:00 pm). Boxplots interpretation for the aggregated data per daytype enables similar understanding for the distributional characteristics of the daytype grouping – longer travel times at weekdays with larger differentiations among paths and same median more or less for Saturdays-Sundays (evidently lower median than weekdays) with different however distribution (lower distribution for Sunday).

One can mine valuable information from the above figure (Fig. 4), comparative as well as per each path separately, however, for the current paper we refer to specific remarks of high interest.

Being paths of more or less similar characteristics, the observed travel time distributions (mix/max) are similar for the majority of paths per each daytype (path 107 that has the bigger length, also presents similar distribution, fact that is counterbalanced from the existence of 4 operational lanes – more lanes than of the rest paths).

Path 121, the main central path of the city serving main commercial land uses with high taxi use, seems to have a significantly different reaction from the rest paths in all daytypes (Fig. 4). Although not being explained by its geometry (by its length), path 121 presents the max travel times during the day (explained by the operation and the role of the path serving the central areas' land uses and trip purposes and being the final destination of vertical axis serving the trips with destination the city center). Furthermore, path 121 presents the max travel time at late Saturday nights and the early hours of Sundays, a situation that was usual (commonly met) years ago when the effects of the economic crisis were not yet visible (prior to 2012). Big travel times (therefore, low speeds of around 10 km/h) are met also in weekdays during the morning peak (09:00–10:00) and the noon (13:00)' (Fig. 5).

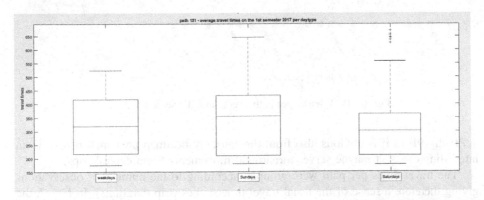

Fig. 5. Average travel time boxplots in path121 per daytype, 1st semester 2017

Timeseries of Paths 7 and 121, being successive parts of the same axis, present similar (and of course expected) shape. The same result seems to be verified also for the rest paths that are successive parts of same axis (with some differentiations as described below).

For weekdays, other paths presenting notable differences compared to the rest, are paths 192 and 104. As for the first, it is a main path with direction towards Eastern Thessaloniki, that serves also the turnaround movements of the city centre (peak hour: 19:00 serving probably work/education to home activities). Path 104 from the other side, being the last leg of the main horizontal axis of the city, Egnatia/Karamanli, that is linked to the highway Thessaloniki – Chalkidiki (serving eastern Thessaloniki and regional connectivity to Chalkidiki Prefecture), presents high travel times at noon (15:00) that verifies the general picture of traffic conditions.

Interesting results arise from the peak hours per path on weekdays (Fig. 6) – the majority of paths with direction towards the centre present peak at 9:00–10:00 while on the opposite direction during the afternoon (15:00–16:00).

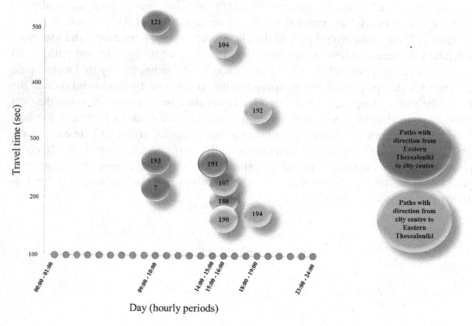

Fig. 6. Peak hours per path, weekdays, 1st semester 2017

Path 191, as it is obvious also from the causality heatmap presented below, is an interesting case – it maybe serves turnaround movements (intra central trips).

The travel times per path were grouped according to the number of observations giving therefore a sense of the usual travel time frames (trip durations). In Fig. 7, the (hereinafter called) 'abnormal' values are highlighted in red.

Looking at the time of the day where abnormal path travel times are observed, the majority of abnormal travel times are found at noon hours (12:00–16:00) while for path 121, abnormal observations are also apparent at night hours (22:00–06:00), an expected fact also from Fig. 4 (higher travel times at path 121 on Saturdays nights). We can say therefore that abnormal observations can reveal peaks or special events (e.g. Fig. 7, 5% of

Path	travel time (sec) / grouped data					total number of taxi that crossed the path at the 1st semester of 2017	taxis per day (on average)
	0-299	300-599	600-899	900-1199	1200-1499		
7	99%	1%	0%	0%	0%	17155	95
104	67%	29%	3%	0%	0%	16562	92
107	99%	0%	0%	0%	0%	17058	94
121	56%	38%	5%	1%	0%	17228	95
188	99%	0%	0%	0%	0%	17124	95
189	93%	5%	1%	0%	0%	17076	94
190	98%	0%	0%	0%	0%	17324	96
191	95%	4%	0%	0%	0%	17270	95
192	89%	9%	1%	0%	0%	17278	95
193	96%	3%	0%	0%	0%	17265	95
194	97%	1%	0%	0%	0%	16983	94

Fig. 7. Taxi (number and percentage) per defined travel time periods, 1st semester 2017 (Color figure online)

the observations for path 121 show travel times of 600–899 s while usually path's crossing duration is lower than 10 min, therefore a special event could have led to such delay).

4.2 Correlation Knowledge Graphs

The correlation analysis was done for typical weekdays, days when the most massive trips (for work and school related activities) take place. Values (correlations) presented in Fig. 8 are normalized. The higher the value the more intense the correlation among the paths.

	path7	path104	path107	path121	path188	path190	path191	path192	path193	path194
path7	1	0,6764	0,7065	0,914	0,7812	0,7061	0,861	0,5905	0,8563	0,6391
path104		1	0,8858	0,7433	0,8422	0,9186	0,9093	0,8538	0,8384	0,8911
path107			1	0,7845	0,7818	0,9425	0,8222	0,8672	0,7827	0,8659
path121				1	0,8056	0,7375	0,915	0,7481	0,9493	0,7526
path188					1	0,8143	0,871	0,6125	0,8458	0,6994
path190						1	0,8459	0,8684	0,7828	0,8636
path191							1	0,6344	0,943	0,7868
path192								1	0,501	0,9047
path193									1	0,6738
path194										1

	path7	path104	path107	path121	path188	path190	path191	path192	path193	path194
path7	0	0	9	0	-2	2	1	23	0	3
path104		0	2	-1	0	1	0	4	0	0
path107			0	-9	-9	-1	-2	1	-9	-3
path121				0	-2	7	1	20	0	12
path188					0	5	2	2	3	1
path190						0	-2	1	-3	0
path191							0	0	0	0
path192								0	0	-1
path193									0	0
path194										0

Fig. 8. Path's correlation (i) and respective time lag (ii), weekdays, 1st semester 2017

The correlation analysis of travel times per path correlations demonstrates the existence of strong road paths- travel time correlations. There is a stronger association between segments (consecutive or not) in the same direction - this is also justified by the time lag as in most cases it is a simultaneous influence or a 15-min difference. The opposite directions are theoretically influenced over a longer period of time (over 6 h - $\sim$ = 24 days - as it is a weekday with the majority of traveling to work and education) - a fact that is visible from time lags between path7/path121 (unidirectional axes with direction to the West) and path 192 (direction to the East).

High correlation accompanied with high causalities (see Sect. 4.3) can be explained as evidence for traffic management interventions; i.e. when we have two road paths x, y highly correlated (with causality existence) when something happens to road path x, we know that the effect will be seem after z seconds in path y, therefore this knowledge could be a basic component for a decision support system in traffic management.

4.3 Causality Knowledge Graphs

As described above, for the current work the authors employed the Conditional Granger Causality from the 'Multivariate Granger Causality Toolbox' (MVGC) [26]. The use of Akaike or Bayesian information criteria seemed to be very significant for results interpretation. Values also in causality analysis (Fig. 9) are normalized. The causality graph is read from column to row (e.g. changes at path 121 seems to highly influence path 7 – value 0.96).

	path7	path121	path107	path104	path188	path190	path191	path192	path193	path194
path7	NaN	0,96	0,66	0,66	0,56	0,67	0,59	0,53	0,99	0,74
path121	0,4	NaN	0,44	0,65	0,81	0,52	0,39	0,49	0,48	0,81
path107	0,58	0,1	NaN	0,16	0,19	0,04	0	0,31	0,14	0,33
path104	0,48	0,59	0,75	NaN	0,51	0,93	0,48	0,75	0,72	1
path188	0,41	0,27	0,61	0,07	NaN	0,42	0,3	0	0,51	0,54
path190	0,47	0,38	0,91	0,2	0,34	NaN	0,53	0,35	0,71	0,84
path191	0,55	0,77	0,77	0,26	0,86	0,45	NaN	0,69	0,47	0,92
path192	0,29	0,21	0,14	0,17	0,42	0,14	0,47	NaN	0,18	0,24
path193	0,6	0,34	0,64	0,38	0,64	0,66	0,47	0,81	NaN	0,93
path194	0,41	0,7	0,67	0,76	0,96	0,79	0,67	0,56	0,63	NaN

Fig. 9. Granger causality results applying BIC criterion, weekdays, 1st semester 2017 involving all examined paths

The use of Akaike (AIC) or Bayesian (BIC) information criteria concluded that for paths with mixed directions, BIC can better explain the causalities among paths (for same direction paths, the results were the same with AIC and BIC).

Comparing correlation and causality results, the expected was verified – subparts of the same axis or paths with same direction are related (e.g. 121-7, 104-194, 107-190). Interesting is however, the direction of causality – i.e. although path 7 precedes path 121, path 121 generates the causality that can be explained from the fact that it is a main generation pole of traffic (city centre). Very interesting is also the role of path 191 (indeed the indication can also be born from Fig. 6) – although being a path that crosses city centre and represents a gate to Western Thessaloniki, it is strongly correlated (with causality of 0.77) and it seems to be influenced by path 121 (the central path) while from the other side does not influence its successor path, path 193 – it seems that end of path 191 acts as a breakpoint of trips (final destination city centre).

4.4 Concluding Remarks

Concluding, it can be supported that correlation along with causality analysis can be an effective interpretation tool for big traffic data. From the above results and based on the knowledge and intuition for the city of Thessaloniki traffic performance, we can suggest that, correlation and Granger causality analysis, that has been widely used in econometric models and subsequently in many other fields, seem to reach notable results also in transportation planning field. Therefore, with the necessary programming steps, a decision support system giving indications for traffic performance at related road paths can be structured. Furthermore and given the existence of multisource data, data fusion techniques can be built upon correlation – causality analysis results.

Acknowledgment. The authors wish to acknowledge the Hellenic Institute of Transport for the access to the traffic data.

References

1. Cai, L., Zhu, Y.: The challenges of data quality and data quality assessment in the big data era. Data Sci. J. **14**, 1–10 (2015). https://doi.org/10.5334/dsj-2015-002
2. Hall, D.L., McMullen, S.A.H.: Mathematical Techniques in Multisensor Data Fusion. Artech House, Norwood (2004). ISBN 1580533353
3. Zhang, L., et al.: Visual analytics for the big data era – a comparative review of state-of-the-art commercial systems. In: 2012 IEEE Conference on Visual Analytics Science and Technology (VAST), pp. 173–182 (2012)
4. Antoniou, C., Balakrishna, R., Koutsopoulos, H.N.: A synthesis of emerging data collection technologies and their impact on traffic management applications. Eur. Transp. Res. Rev. **3**, 139–148 (2011). https://doi.org/10.1007/s12544-011-0058-1
5. Leduc, G.: Road Traffic Data: Collection Methods and Applications. JRC 47967 – Joint Research Centre – Institute for Prospective Technological Studies. Office for Official Publications of the European Communities, Luxembourg (2008)
6. Myrovali, G., Tsaples, G., Morfoulaki, M., Aifadopoulou, G., Papathanasiou, J.: An interactive learning environment based on system dynamics methodology for sustainable mobility challenges communication & citizens' engagement. In: Dargam, F., Delias, P., Linden, I., Mareschal, B. (eds.) ICDSST 2018. LNBIP, vol. 313, pp. 88–99. Springer, Cham (2018). https://doi.org/10.1007/978-3-319-90315-6_8
7. Patire, A.D., Wright, M., Prodhomme, B., Bayen, A.M.: How much GPS data do we need? Transp. Res. Part C **58**, 325–342 (2015)
8. Hall, D.L., Llinas, J.: An introduction to multisensor data fusion. Proc. IEEE **85**, 6–23 (1997)
9. Varshney, P.K.: Multisensor data fusion. Electron. Commun. Eng. J. **9**, 245–253 (1997)

10. Faouzi, N.-E.E., Leung, H., Kurian, A.: Data fusion in intelligent transportation systems: progress and challenges a survey. Inform. Fusion **12**, 4–10 (2011). Special Issue on Intelligent Transportation Systems
11. Ranjan, R., et al.: City data fusion: sensor data fusion in the Internet of Things. Int. J. Distrib. Syst. Technol. **7**(1), 15–36 (2016)
12. Qing, O.: Fusing Heterogeneous Traffic Data: Parsimonious Approaches Using Data-Data Consistency. T2011/5, TRAIL Thesis Series, The Netherlands (2011)
13. Khaleghi, B., Khamis, A., Karray, F.O., Razavi, S.N.: Multisensor data fusion: a review of the state-of-the-art. Inf. Fusion **14**(1), 28–44 (2013)
14. Mitsakis, E., Stamos, I., Salanova Grau, J.M., Chrysochoou, E., Iordanopoulos, P., Aifadopoulou, G.: Urban mobility indicators for Thessaloniki. J. Traffic Logistics Eng. **1**(2), 148–152 (2013)
15. Stamos, I., Salanova Grau, J.M., Mitsakis, E.: Modeling Effects of Precipitation on Vehicle Speed: Floating-Car Data Approach. TRB 2016 Annual Meeting (2016)
16. Chien, S.I.J., Kuchipudi, C.M.: Dynamic travel time prediction with real-time and historic data. J. Transp. Eng. **129**(6), 608–616 (2003)
17. Mitsakis, E., Salanova Grau, J.M., Chrysohoou, E., Aifadopoulou, G.: A robust method for real time estimation of travel times for dense urban road networks using point-to-point detectors. Transport **30**(3), 264–272 (2015). https://doi.org/10.3846/16484142.2015.1078845
18. Charakopoulos, A.K., Katsouli, G.A., Karakasidis, T.E.: Dynamics and causalities of atmospheric and oceanic data identified by complex networks and Granger causality analysis. Physica A **495**, 436–453 (2018)
19. Gao, Z.K., Small, M., Kurths, J.: Complex network analysis of time series. Europhy. Lett. **116**(5), 50001 (2016). https://doi.org/10.1209/0295-5075/116/50001
20. Chatfield, C.: Time-Series Forecasting. Chapman & Hall/CRC, Boca Raton (2000). ISBN 1-58488-063-5
21. STAT 510 – Applied Time Series Analysis, Lesson 8: Regression with ARIMA errors, Cross correlation functions, and Relationships between 2 Time Series, 8.2 Cross Correlation Functions and Lagged Regressions. https://newonlinecourses.science.psu.edu/stat510/node/74/
22. Granger, C.W.J.: Investigating causal relations by econometric models and cross-spectral methods. Econometrica **37**(3), 424–438 (1969). https://doi.org/10.2307/1912791
23. Roebroeck, A., Formisano, E., Goebel, R.: Mapping directed influence over the brain using Granger causality and fMRI. NeuroImage **25**(1), 230–242 (2005). https://doi.org/10.1016/j.neuroimage.2004.11.017
24. Attanasio, A.: Testing for linear Granger causality from natural/anthropogenic forcings to global temperature anomalies. Theoret. Appl. Climatol. **110**, 281–289 (2012)
25. Charakopoulos, A.K., Karakasidis, T.E., Liakopoulos, A.: Spatiotemporal analysis of seawatch buoy meteorological observations. Environ. Process. **2**(1), 23–39 (2015)
26. Barnett, L., Seth, A.K.: The MVGC multivariate Granger causality toolbox: a new approach to Granger-causal inference. J. Neurosci. Methods **223**, 50–68 (2014)

Value-Chain Wide Food Waste Management: A Systematic Literature Review

Guoqing Zhao[1(✉)], Shaofeng Liu[1], Huilan Chen[1], Carmen Lopez[1],
Jorge Hernandez[2], Cécile Guyon[3], Rina Iannacone[4],
Nicola Calabrese[5], Hervé Panetto[6], Janusz Kacprzyk[7],
and MME Alemany[8]

[1] University of Plymouth, Drake Circus, Plymouth PL4 8AA, UK
{guoqing.zhao,shaofeng.liu,huilan.chen,
carmen.lopez}@plymouth.ac.uk
[2] University of Liverpool, Liverpool L69 3BX, UK
J.E.Hernandez@liverpool.ac.uk
[3] Bretagne Développement Innovation,
1bis Rue de Fougères, 35510 Cesson-Sévigné, France
c.guyon@bdi.fr
[4] ALSIA, Viale Carlo Levi, 75100 Matera, MT, Italy
rina.iannacone@gmail.com
[5] CNR-ISPA, Via Amendola, 122/O, 70126 Bari BA, Puglia, Italy
nicola.calabrese@ispa.cnr.it
[6] University of Lorraine, 34 cours Léopold, CS 25233,
54052 Nancy Cedex, France
herve.panetto@univ-lorraine.fr
[7] IBSPAN, Newelska 6, 01-447 Warsaw, Poland
janusz.kacprzyk@ibspan.waw.pl
[8] Universitat Politècnica de València,
Camino de Vera, s/n, 46022 Valencia, Spain
marvea@omp.upv.es

Abstract. The agriculture value chain, from farm to fork, has received enormous attention because of its key role in achieving United Nations Global Challenges Goals. Food waste occurs in many different forms and at all stages of the food value chain, it has become a worldwide issue that requires urgent actions. However, the management of food waste has been traditionally segmented and in an isolated manner. This paper reviews existing work that has been done on food waste management in literature by taking a holistic approach, in order to identify the causes of food waste, food waste prevention strategies, and elicit recommendations for future work. A five step systematic literature review has been adopted for a thorough examination of the existing research on the topic and new insights have been obtained. The findings suggest that the main sources of food waste include food overproduction and surplus, food waste caused by processing, logistical inconsistencies, and households. Main food waste prevention strategies have been revealed in this paper include policy solutions, packaging solutions, date-labelling solutions, logistics solutions, changing consumers' behaviours, and reuse and redistribution solutions. Future research directions such as using value chain models to reduce food waste and

© Springer Nature Switzerland AG 2019
P. S. A. Freitas et al. (Eds.): EmC-ICDSST 2019, LNBIP 348, pp. 41–54, 2019.
https://doi.org/10.1007/978-3-030-18819-1_4

forecasting food waste have been identified in this paper. This study makes a contribution to the extant literature in the field of food waste management by discovering main causes of food waste in the value chain and eliciting prevention strategies that can be used to reduce/eliminate relevant food waste.

Keywords: Systematic literature review · Food waste generation ·
Cause of waste · Food waste prevention strategies

1 Introduction

In 2011 the Food and Agriculture Organisation (FAO) of the United Nations estimated that roughly one third of the worldwide food produced for human consumption is lost or wasted from the initial stage of production to the final stage of consumption, which amounts to about 1.3 billion tons per year [1]. According to the European Commission [2], food waste has been classified into three different categories: (1) "food losses: food products lost during the production phase; (2) unavoidable food waste: referring to food products lost during the consumption phase (banana peels, fruit cores, etc.); (3) avoidable food waste: products that could have been eaten, but were lost during consumption phase". Food losses typically but not exclusively happen at the production, postharvest and processing stage of the food value chain [3], whereas food waste always takes place at the final stage of food value chain such as during retail and consumption stage [4].

To reduce food waste in food value chains, food waste management has received great attention in recent years. [5] investigated the root causes of food waste from the retail and store operations perspective. Their findings indicate that undesirable customer behaviour and erratic demand, inefficient store operations and replenishment policies, and elevated product requirements of both retail organisations and customers are the root causes of food waste. [6] found that investigating the causes of food waste deepens the understanding in food waste and helps to design food waste prevention strategies. However, most of studies concerned with food waste management exclusively on a single stage of food value chain, e.g. either focusing on retailer [5] or consumer [6], not from a whole food value chain perspective. Compared with existing literature, main contributions of this paper include the provision of elicitation of value chain wide (covering all stages from food production, manufacturing through logistics/distribution and retailing to consumption) food waste sources, a holistic view of food waste prevention challenges and strategies, key research gaps, and recommendation for future research directions.

In order to have a deep understanding on food waste from whole food value chain perspective, we aim to address two key questions: (1) What are the main causes of food waste in food value chains? (2) What are the key food waste prevention strategies that can be used in food value chains? In order to answer the above questions, the remainder of this paper is organised as follows. The process of systematic literature review (SLR) is described in Sect. 2. Then, the thematic analysis about causes of food waste and the possible food waste prevention strategies are outlined in Sect. 3. In Sect. 4, research gaps are identified from the SLR and future research directions are proposed.

Finally, conclusions and the main connections between food waste generation and food waste prevention strategies are discussed in Sect. 5.

2 Research Methodology: Systematic Literature Review

A SLR is an explicit and methodical way to gain a deep understanding on a given topic to inform academics and practitioners, which is a systematic, reproducible and rigorous approach to identify, synthesize, interpret the best evidence from previous literatures [7]. Unlike traditional narrative reviews, a SLR follows a set of strict guidelines to search for the most relevant literatures in a comprehensive and transparent way, and apply an evidence-based explanation with specific criteria of the body of previous literatures [8]. This paper follows the five phases of SLR methodology proposed by [7]: (1) Formulating the research questions; (2) Locating the study; (3) Study selection and evaluation; (4) Analysis and synthesis; (5) Reporting and using the results. These phases are described in detail in the following sub-sections.

2.1 Formulating the Research Questions

The first stage for conducting SLR is to formulate research questions, which should clearly focus on the causes of food waste as well as the food waste prevention strategies used in food value chains. In our paper, the research questions were formulated based on the following process: (1) Brainstorming all types of food waste in food value chains; (2) Consult with experts, academics in the agri-food industry, to identify root causes of food waste; (3) Examine possible strategies to reduce food waste. Thus, two well specified, informative and clearly focused research questions were formulated in the following: (1) What are the main causes of food waste in food value chains? (2) What are the key food waste prevention strategies that can be used in food value chains?

2.2 Locating Study

The purpose of searching through different databases is to build a comprehensive list of core contributions pertinent to the review questions [7]. Therefore, Web of Science and Business Source Complete were selected for literature search as they include the global major journals and conference proceedings especially in social science [9]. In our review, defined keywords were combined through Boolean connectors in order to constitute different search strings to be searched in the title, keywords and abstract. Since the focus of our study is the causes of food waste and food waste prevention strategies, the strings were specially designed in order to select the most relevant papers, identifying overlaps between food waste generation and food waste prevention in general. Search strings partly include "food waste" AND "value chain", "food waste prevention strategies" AND "value chain". As an example, a full list of keywords and search strings is shown in Table 1. In addition to the keywords and search strings, relevant literatures were also identified through cross-referencing based on previous literatures on the food waste management to include potential papers that had not been selected from the above-mentioned databases. Furthermore, recommendations from practitioners and academics in the relevant field were also adopted to include papers [10]. Thus, the electronic search process resulted in the identification of 788 articles from the Web of Science and 83 articles from Business Source complete.

Table 1. Search strings of food waste management

Keywords	Food waste, Food waste management, Food waste generation, Causes of food waste, Food waste prevention, Food waste inhibition, Value chain, Supply chain, Supply network,
Databases	Web of Science, Business Source Complete
Search strings	"Food waste" AND "(Value chain OR Supply chain OR Supply network)"; "Food waste prevention" AND "(Value chain OR Supply chain OR Supply network)"; "Food waste generation" AND "(Value chain OR Supply Chain OR Supply network)"; "Causes of food waste" AND "(Value chain OR Supply chain OR Supply network)"; "Food waste management" AND "(Value chain OR Supply chain OR Supply network)"; "Food waste inhibition" AND "(Value chain OR Supply chain OR Supply network)"

2.3   Study Selection and Evaluation

The selected keywords were used to search in the above-mentioned two databases with a time limitation from 1980 to 2017, which resulted in a preliminary sample of 871 contributions. The year 1980 was selected as the starting point because the emergence of food waste studies is around this time [6]. Furthermore, the review was limited to papers that published in international peer-reviewed journals in English because peer-reviewed journals are considered to have better quality than non-peer-reviewed journals. Furthermore, articles could help to answer the formulated questions, provide usable methods to develop this study, and offer necessary data or information to this study, were all included to process in-depth analysis. In order to reduce any subjective bias and enhance validity, each paper was judged by three different authors. Through carefully reading the title, abstract, introduction, and conclusion of each paper, in total 871 papers were screened at this stage. Papers were not relevant to the topic or duplicates or non-relevant to the research questions or limits to access were excluded. As the outcome of this process, 140 papers were selected. Then, all the selected articles were read entirely, again by the same three authors independently. By crossing-referencing and consulting with experts in the relevant field, a further 7 articles were identified. Finally, 147 papers were prepared to do in-depth analysis.

2.4   Analysis and Synthesis

The objective of this phase is to analyse and synthesise the selected 147 papers so as to develop new knowledge about the food waste generation and food waste prevention that were not visible through reading each of paper individually. In-depth analysis was conducted through recording the summary of each of the papers into the spreadsheet of Microsoft Excel based on the following criteria: (1) Selected papers were classified based on their general details such as year of publication, journal title, authors' nationality, and methodology adopted; (2) Then, each of paper were classified according to their focus of each study and key issue investigated. Synthesis was achieved through building connection between the themes in each of the paper. Four different approaches such as aggregation, integration, interpretation, and explanation are used in the synthesis process [11]. Through recursively considering the topics found in the literature and discussing the evidence that had emerged from literature with other authors, two key themes were emerged: (1) Food waste were generated in different stages of food value chain from production, processing, transportation, and finally by consumers; (2) Different food waste prevention strategies were formulated to reduce food waste in the food value chain.

2.5   Reporting and Using the Results

This stage of SLR provides the findings from the selected papers, the relationships between each other and the extent of what is known and what is not known about the research questions [7]. Furthermore, findings can help to generate new knowledge and provide recommendations for future research.

3   Findings

A succinct analysis of the basic characteristics of the selected 147 papers shown that a majority of papers were published between 2010 and 2017, providing evidence of academic interest on food waste management has steadily increased. Furthermore, the selected 147 papers were published in 52 interdisciplinary academic journals. More than one third of papers were published in *Waste Management, Journal of Cleaner Production* and *Resources Conservation and Recycling*. The diversity of the journals' research themes (i.e. environment science, food research, logistics research, sociological research) is the evidence of the multidisciplinary nature of the research and the increasing attention is attracted from various research disciplines.

In the following sub-sections, the causes of food waste and food prevention strategies are outlined and discussed in the following, each of which will deepen our understandings and elicit current trends on food waste management. First of all, the causes of food waste were analysed and categorized into different themes. Secondly, food waste prevention strategies and related managerial practices are examined.

3.1   Thematic Analysis: Causes of Food Waste in the Value Chain

In order to provide edible products for human consumption, food has to pass multiple stages along the value chain. It begins with producing of food from the agricultural sector, followed by processing and packaging food in the manufacturing industry, and then distribution to markets through transportation systems, and finally to sell to consumers through retailing system. Food waste occurs in all stages of the food value chain, four main areas of the causes of food waste were identified by recursively considering the topics found in the literature: food over-production and surplus, food wasted during food processing, product deterioration and spoilage during logistics operations, and food waste caused by households.

Food Overproduction and Surplus. An important perspective of food waste generation is over-production. The need for over-production mainly stems from trade demands: the uncertainty to meet quality requirements and to be burdened by contractual penalties or product take-back obligations may cause a higher production volume [4]. Insufficient communication and reluctance to share information between food value chain members are the main reasons for poor forecasting [14]. After conducting various interviews with representatives from 13 German food processing companies, [12] revealed that producers rely on previous experience and past sales numbers to forecast production volume. Their research also indicate that in order to fulfil customers requirements, over-production is tolerated to some extent especially for some fresh bakery products. In this way, consumers' anticipation of permanently full shelves help to foster over-production, and retailers are forced to find the right balance between a maximum on-shelf-availability and a minimum of unsold food surplus [15]. After investigating 83 firms in the sector of food manufacturing and retailing, the primary source of surplus food is revealed. Five main reasons for the generation of surplus food in the food manufacturing industry are: products that have reached the internal sell-by date, product non-compliance with commercial standards, non-compliance of product packing with required standards, product refusals and return of unsold products. Besides the surplus food generated in the food manufacturing industry, four main reasons for the generation of surplus food in the food retail sector are: it has reached the sell-by date, product packing and product itself do not comply with required standards, and product returns [16].

Food Waste Caused by Processing. According to the data from [17], approximately a quarter of food waste was generated by food manufacturing-179 kg/head. This was also confirmed by [18], their research indicated that food waste occurs at the upstream stages of value chains, and up to 12% for fruit, vegetables and potatoes were wasted from production to processing. [12] proposed that food waste generates mainly from five dimensions during the stage of food processing: (1) Internal requirements such as resource supply and raw material quality; (2) External requirements such as processing safety and quality requirements; (3) Intentional losses such as cleaning losses and samples for analysis; (4) Unintentional losses such as power blackouts and equipment defects; (5) Un-utilised by-products.

Food Waste Caused by Logistics. In different stages of food value chain, logistics service providers assume the responsibility for mechanical product damages in transporting products from one stage to the other. These damages can cause different consequences, for example, products damage or contaminated in the transportation process that leads to rejection by the recipient [12]. In the case of raw or minimally processed foods, such as most tropical and subtropical fruits, a high fraction of food waste was contributed by insufficient cold chain utilisation [19]. In other circumstances, it is easy to cause impacts, compressions and abrasions on fresh products during the transportation process, if boxes are stacked incorrectly or if packages are used in improper size [20]. Other reasons for damaged packaging included vibrations during shipping process, a load-factor in outer packaging, overhang on pallets and that moisture weakened the courier paper [21].

Food Waste Caused by Households. Most food waste is caused by downstream of the food value chain, especially in the interaction between retail, food service and the consumers as well as in consumer's home [22]. [4] defined the household food waste as "sources of food and drinks that are consumed within the home include retail and contributions from home-grown food and takeaways". According to the data from FAO of the United Nations, less than 10% of the wastage occurs at the distribution level of Europe, but more than 30% of wastage are caused by consumers [1]. The potential reasons for food waste in households were listed in the following Table 2:

Table 2. Potential reasons for food waste generation in households

Potential reasons for food waste generation		Authors
Planning	1. Lack of planning of food shopping and meals	[23]
	2. Inadequate communication between household members	[24]
Shopping	1. Differences in taste 2. Time constraints 3. Oversized packaging	[24]
Storing	1. Improper and unsystematic storage practices	[25]
Cooking	1. Preference of convenience food	[26]
	2. Over-preparation of food	[13]
Eating	1. Unpredictable eating patterns; 2. Eating-out in restaurants; 3. Large plate sizes	[27]
Managing leftovers	1. Wish for variety in meals	[13]
	2. Lack of knowledge about leftovers	
Assessing edibility	1. Confusion of date-labels	[27]
	2. Lack of knowledge about shelf-life food and how to extend it	
Dsiposal/Redistribution	1. Lack of social acceptance of food sharing	[28]
		[29]

A growing body of literature has investigated food-practices and routines in the context of food waste generation. Many causes of food waste have direct relationship with consumers, for example, lacking planning routines and cooking capabilities, misunderstanding of date-labelling, demanding high quality level of food that they want to buy, and avoiding potential risks [30, 31]. [30] considered the effect of psycho-social factors, food-related routines, household perceived capabilities and socio-demographic on food waste, the results indicated that perceived behavioural control, routines related to shopping and reuse of leftovers are the main drivers of food waste, while planning routines contribute indirectly. [32] found that the size of plates have a positive correlation with the amount of food wasted, and larger plates induce people to eat more. [33] found that people who like to eat ready-made meals and restaurant take away will waste more edible food than others.

3.2 Thematic Analysis: Food Waste Prevention Strategies

Appropriate food waste management is recognised as essential for sustainable development [34]. Reducing food waste throughout the food value chain should be considered as an effective solution to increase the income of food value chain members and to improve food security for low-income consumers [35].

Policy Solutions. Regulatory approaches, such as laws and standards, mandatory management plans, restrictions or covenants that can help to achieve the target of reducing waste and induce consumers to form good behaviour. [36] proposed that one possible way to reduce food waste is to eliminate unnecessary food safety standards that lead to high food waste rates. [37] hold the same view that well-defined regulations seem to be a good weapon to fight with household food waste generation. Incentive-based regulatory approach such as economic incentives (fees, taxes and subsides) is another way that can be used to reduce food waste. [38] proposed that financial instruments have significant status in shifting consumer behaviours towards more sustainable food practices. A cross-country analysis has been conducted by [37] after collecting data from 44 countries with various income levels, their analysis results indicated that well-defined regulations, policies and strategies have a positive effect in mitigating household food waste generation. The most widespread tools used for reducing and preventing food waste are information campaigns [39], which have been implemented all over Europe to improve consumer's knowledge raise awareness about food waste prevention.

Packaging Solutions. Packaging plays an important role in containing and protecting food as it moves through the value chain from producers to final consumers. Food waste can be reduced through the use of packaging that improves the lifespan of food, ventilation and temperature control [40]. In order to extend the lifespan of food, innovation and evolution in intelligent packaging with improved protection, communication, convenience, and containment are slowly entering the market [41]. Different technologies aimed at extending the shelf-life of food have been developed, such as multiple-layer barrier packaging, modified atmosphere packaging, oxygen scavengers and aseptic packaging. The most widespread used technology – oxygen scavengers, removing oxygen slows the degradation process and extends the shelf life of the food [40].

Date-Labelling Solutions. Date-labelling plays a vital role in informing both retailers and consumers about how long a food will remain edible, safe and of sufficient quality makes it a prime site for the identification of, intervention in, food waste [42]. As [39] proposed that two important reasons that can cause avoidable consumer food waste are: (1) lack of knowledge about the meanings of date labels; (2) confusion about the difference between expiry date and the date of minimum durability. In order to reduce confusion among consumers about expiry dates and the date of minimum durability, a big potential for reducing food waste lies in improving labels for pre-packed food products, such as removing sell-by-date labels or reducing the indicating date of minimum durability used in some products [36]. Furthermore, advanced technology used in the labelling may help to reduce consumer food waste, such as Time-Temperature-Indicator device can help to identify low quality and potentially unsafe food [43].

Logistics Solutions.   [44] proposed that efficient information sharing among value chain partners and adaption of material flows, which have a positive effect on reducing food waste in the food value chain. [45] found that food waste can be reduced through taking shelf-life losses into consideration, if the result of standard cost parameters in poor product quality and large amounts of product waste were determined. Meanwhile, other researchers on food logistics have investigated the different areas of logistics that can help to reduce the food waste, for example, [46] conducted a simulation through using the information on the quality of products that provided by intelligent packaging, the results indicated that quality-controlled logistics can substantially reduce food waste. Further, after conducting 19 semi-structured interviews and four site visits in the Swedish food value chains, [21] concluded that collaborative forecasting, division of lead time, low level of safety stock, packaging development, visualising damaged packaging, product group revisions, price reductions, make-to-order flows and measure of service levels, all these logistics activities can have a positive effect on food waste reduction.

Reuse and Redistribution Solutions.   [47] proposed different methods that can be used to reduce food waste. First, surplus food can be used to donate to non-profit organizations to help homeless people or used to sponsor specific organisations. Second, food companies can choose to distribute surplus products within their organisations or sales in the secondary market. Third, based on the type of products and the reason why the food waste generated, companies can remanufacture or repackage it. Finally, surplus food can be sold with promotions and discounts, or to sell to companies that produce animal feed or fertilizers.

Changing Consumers' Behaviours.   A growing body of literature has investigated how to prevent food waste through changing consumers' behaviours. [13] proposed that careful planning of grocery shopping is an effective tool to prevent food waste. After collecting survey data from 1062 Danish respondents, [30] concluded that providing advice on how to deal with food-related activities at home through booklets or other such communication means, which have a positive effect on changing consumers' food waste behaviour. In addition, their research also indicates that participation in cooking courses or implementing household economics education campaigns are also effective in changing consumers' food waste behaviour. The possible food waste prevention measures related changing consumers' food waste behaviour were listed in the following Table 3:

Table 3. Possible food waste prevention measures

Possible prevention measures		Authors
Planning	1. Using a shopping list	[23]
	2. Communication with household members...	[24]
Shopping	1. Increase shopping frequency	[24]
Storing	1. Systematically storing and categorizing	[25]
	2. Freezing of food...	
Cooking	1. Greater frequency of cooking	[13]
	2. Better estimation of portion size...	[26]
Eating	1. Household members with special diets	[27]
Managing leftovers	1. Reusing leftovers	[13]
		[30]
Assessing edibility	1. More nuanced assessments of food edibility	[27]
Disposal/Redistribution	1. Giver food waste to pets	[28]
	2. Recycling and composting	[29]
Information and knowledge sharing	1. Implementing household economics education campaigns	[30]
	2. Food waste related booklets...	

4 Research Gaps and Future Directions

The management of food waste has attracted the attention of researchers and practitioners over the last forty years and the focus has been on the causes of food waste, the reduction of the sources of the food waste from policy perspectives [37], packaging perspectives [40], date-labelling perspectives [39], logistics perspectives [44], changing consumers' behaviours perspective, and reuse and redistribution perspectives [47]. The overview of the causes of food waste and food waste prevention strategies identified in this paper were demonstrated in Fig. 1. In addition, the use of the advanced technologies that monitors production along the food value chain [48], the trends that using technologies to process the food waste [49], and the cost of food waste along the value chain have also been identified in the literature review process. Based on the review, the following research gaps have been identified and future research directions can be proposed: (1) Few researchers have explored how to forecast food waste in food value chains. Future research should investigate how to use holistic-adaptive forecasting methods to forecast food waste; (2) Little research considering how to reduce food waste from a logistics perspective involving producers/farmers. Therefore, future research should investigate how producers/farmers can participate in logistics solutions, involving efficient information sharing to reduce food waste; (3) Although different authors have proposed different solutions to reducing food waste in food value chains, these authors just investigated how to reduce food waste from one perspective. Future work is required that tests and assesses the effectiveness and impact of different food waste solutions combined together; (4) Little research exploring the impact of consumers' attitudes to reducing food waste in food value chains. Therefore, conducting semi-structured interviews with different consumers on how to alter consumers' attitudes on suboptimal food can be as an important future research direction to reduce food waste; (5) Furthermore, little research has been conducted on how to reduce food waste through using value chain models. Future research should investigate new value chain coordination models, planning and replenishment strategies that can be used to reduce food waste.

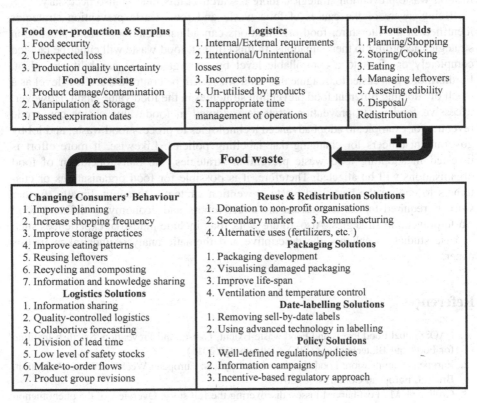

Food over-production & Surplus
1. Food security
2. Unexpected loss
3. Production quality uncertainty
Food processing
1. Product damage/contamination
2. Manipulation & Storage
3. Passed expiration dates

Logistics
1. Internal/External requirements
2. Intentional/Unintentional losses
3. Incorrect topping
4. Un-utilised by products
5. Inappropriate time management of operations

Households
1. Planning/Shopping
2. Storing/Cooking
3. Eating
4. Managing leftovers
5. Assesing edibility
6. Disposal/ Redistribution

Food waste

Changing Consumers' Behaviour
1. Improve planning
2. Increase shopping frequency
3. Improve storage practices
4. Improve eating patterns
5. Reusing leftovers
6. Recycling and composting
7. Information and knowledge sharing
Logistics Solutions
1. Information sharing
2. Quality-controlled logistics
3. Collabortive forecasting
4. Division of lead time
5. Low level of safety stocks
6. Make-to-order flows
7. Product group revisions

Reuse & Redistribution Solutions
1. Donation to non-profit organisations
2. Secondary market   3. Remanufacturing
4. Alternative uses (fertilizers, etc.)
Packaging Solutions
1. Packaging development
2. Visualising damaged packaging
3. Improve life-span
4. Ventilation and temperature control
Date-labelling Solutions
1. Removing sell-by-date labels
2. Using advanced technology in labelling
Policy Solutions
1. Well-defined regulations/policies
2. Information campaigns
3. Incentive-based regulatory approach

Fig. 1. The overview of the causes of food waste and food waste prevention strategies

5 Discussion and Conclusions

In this study, we conducted a systematic literature review to identify the sources of food waste generation and food waste prevention strategies in food value chains. Through this process, we examined in total 147 journal papers. Main causes of food waste generation include food over-production and surplus, food waste caused by manufacturing, food waste caused by logistics and food waste caused by households. At the value chain level, some food waste can be controlled, such as food overproduction and food surplus can be reduced through communication and collaboration with value chain partners to increase the forecasting; food waste caused by manufacturing can also be reduced through production process optimization. However, some food waste caused by consumers is very difficult to control. Therefore, more empirical studies are needed to explore which methods are effective in reducing consumer-related food waste. As for the food waste prevention strategies, food waste can be reduced through developing appropriate policies, changing packaging and improving date-labelling, enhancing collaboration between food value chain partners, and reusing and redistributing of food waste. Although different researchers have conducted various studies in food waste prevention strategies, more research in this field is also necessary.

Being aware of the causes of food waste and food waste prevention strategies identified in the literature, food organisations can adjust their food waste prevention strategies according to their situations. In this context, food waste will either be tackled completely or offset to a controllable level by an organisation capability to adapt. Furthermore, it is easier to imagine that food waste will be controlled at a low level as a result of adopting different food prevention strategies in the food value chain. However, excessive food waste prevention strategies adopted in food value chains can erode revenue, for example, to adopt advanced technologies to process food waste, and lobby government officers for changing date-labelling policies. Likewise, if more effort is invested in adopting food waste prevention strategies, the daily operation of food organisations will be affected. Therefore, it is possible for food organisations or customers to choose the best food waste prevention strategies to adopt. Finally, further work is required to test and assess the effectiveness and economy of different food waste prevention strategies. Also, future work may include industrial experiences such as case studies to enhance the descriptive and thematic analysis undertaken in this paper.

References

1. FAO: Global Food losses and Food waste-Extent, Causes and Prevention. Swedish Institute for Food and Biotechnology (SIK), Gothenburg (2011)
2. European Commission: Food Waste and Its Impacts: European Week for Waste Reduction, Brussel, European Commission (2014)
3. Grolleaud, M.: Post-harvest losses: discovering the full story. Overview of the phenomenon of losses during the post-harvest system. FAO, Rome (2002)
4. Parfitt, J., et al.: Food waste within food supply chains: quantification and potential for change to 2050. Philos. Trans. R. Soc. B: Biol. Sci. **365**, 3065–3081 (2010)
5. Teller, C., et al.: Retail store operations and food waste. J. Clean. Prod. **185**, 981–997 (2018)
6. Schanes, K., et al.: Food waste matters–a systematic review of household food waste practices and their policy implications. J. Clean. Prod. **182**, 978–991 (2018)
7. Denyer, D., Tranfield, T.: The Sage Handbook of Organisational Research Methods. Sage, London (2009)
8. Tranfield, D., Denyer, D., Smart, P.: Towards a methodology for developing evidence-informed management knowledge by means of systematic review. Br. J. Manag. **14**, 207–222 (2003)
9. Durach, C.F., et al.: Antecedents and dimensions of supply chain robustness: a systematic literature review. Int. J. Phys. Distrib. Logist. Manag **45**, 118–137 (2015)
10. Datta, P.: Supply network resilience: a systematic literature review and future research. Int. J. Logist. Manag. **28**, 1387–1424 (2017)

11. Rousseau, D.M., Manning, J., Denyer, D.: Evidence in management and organizational science: assembling the field's full weight of scientific knowledge through syntheses. Acad. Manag. Ann. **2**, 475–515 (2008)
12. Raak, N., et al.: Processing- and product-related causes for food waste and implications for the food supply chain. Waste Manag. **61**, 461–472 (2017)
13. Secondi, L., Principato, L., Laureti, T.: Household food waste behaviour in EU-27 countries, a multilevel analysis. Food Pol. **56**, 25–40 (2015)
14. Mena, C., et al.: The causes of food waste in the supplier-retailer interface: evidences from the UK and Spain. Res. Conserv. Recycl. **55**, 648–658 (2011)
15. Ettouzani, Y., et al.: Examing retail on shelf availability: promotional impact and a call for research. Int. J. Phys. Distrib. Logist. Manag. **42**, 213–243 (2012)
16. Garrone, P., et al.: Surplus food recovery and donation in Italy: the upstream process. Br. Food J. **116**, 1460–1477 (2014)
17. European Commission: EU Guidance to the Commission Regulation (EC) No 450/2009 of May 2009 on Active and Intelligent Materials and Articles Intended to Come into Contact with food. Version 1.0. Health and Consumers Directorate-General of European Commission, Belgium (2011)
18. Redlingshofer, B., et al.: Quantifying food loss during primary production and processing in France. J. Clean. Prod. **164**, 703–714 (2017)
19. Jedermann, R., et al.: Reducing food losses by intelligent food logistics. Philos. Trans. R. Soc. A **372**, 20130302 (2014)
20. Li, Z., Thomas, C.: Quantitative evaluation of mechanical damage to fresh fruits. Trends Food Sci. Technol. **35**, 138–150 (2014)
21. Liljestrand, K.: Logistics solutions for reducing food waste. Int. J. Phys. Distrib. Logist. Manag. **47**, 318–339 (2017)
22. Gobel, C., et al.: Cutting food waste through cooperation along the food supply chain. Sustainability **7**, 1429–1445 (2015)
23. Farr-Wharton, G., et al.: Identifying factors that promote consumer behaviours causing expired domestic food waste. J. Consum. Behav. **13**, 393–402 (2014)
24. Jorissen, J., et al.: Food waste generation at household level: results of a survey among employees of two European research centers in Italy and Germany. Sustainability **7**, 2695–2715 (2015)
25. Martindale, W.: Using consumer surveys to determine food sustainability. Br. Food J. **116**, 1194–1204 (2014)
26. Graham-Rowe, E., et al.: Identifying motivations and barriers to minimising household food waste. Res. Conserv. Recycl. **84**, 15–23 (2014)
27. Parizeau, K., et al.: Household-level dynamics of food waste production and related beliefs, attitudes and behaviours in Guelph, Ontario. Waste Manag. **35**, 207–217 (2015)
28. Wenlock, R., et al.: Household food wastage in Britain. Br. J. Nutri. **43**, 53–70 (1980)
29. Tucker, C., Farrelly, T.: Household food waste: the implications of consumer choice in food from purchase to disposal. Local Environ. **21**, 682–706 (2015)
30. Stacu, V., et al.: Determinants of consumer food waste behaviour: two routes to food waste. Appetite **96**, 7–17 (2016)
31. Waston, M., Meah, A.: Food waste and safety: negotiating conflicting social anxieties into the practices of domestic provisioning. Sociol. Rev. **60**, 102–120 (2013)
32. Wansink, B., Van Ittersum, K.: Portion size me: plate-size induced consumption norms and win-win solutions for educing food intake and waste. J. Exp. Psychol. Appl. **19**, 320–332 (2013)
33. Mallinson, L.J., et al.: Attitudes and behaviour towards convenience food and food waste in the United Kingdom. Appetite **103**, 17–28 (2016)

34. UNEP: Decoupling Natural Resource Use and Environmental Impacts from Economic Growth. United Nations Environment Programme, Paris (2011)
35. Rezaei, M., Liu, B.: Food loss and waste in the food supply chain. FAO, Rome (2017). http://www.fao.org/save-food/news-and-multimedia/news/news-details/en/c/1026569/. Accessed 19 May 2018
36. Schanes, K., et al.: Low carbon lifestyles: a framework to structure consumption strategies and options to reduce carbon footprints. J. Clean. Prod. **139**, 1033–1043 (2016)
37. Chalak, A., et al.: The global economic and regulatory determinants of household food waste generation: a cross-country analysis. Waste Manag. **48**, 418–422 (2016)
38. Reisch, L.A., et al.: Sustainable food consumption: an overview of contemporary issues and policies. Sustain.: Sci. Pract. Pol. **9**, 7–25 (2013)
39. Priefer, C., et al.: Food waste prevention in Europe-a cause-driven approach to identify the most relevant leverage points for action. Res. Conserv. Recycl. **109**, 155–165 (2016)
40. Verghese, K., et al.: Packaging's role in minimizing food loss and waste across the supply chain. Packag. Technol. Sci. **28**, 603–620 (2015)
41. Vanderroost, M., et al.: Intelligent food packaging: the next generation. Trends Food Sci. Technol. **39**, 47–62 (2014)
42. Milne, R.: Arbiters of waste: date labels, the consumer and knowing good, safe food. Sociol. Rev. **60**, 84–101 (2013)
43. Newsome, R., et al.: Applications and perceptions of date labelling of food. Compr. Rev. Food Sci. Food saf. **13**, 745–769 (2014)
44. Kaipia, R., et al.: Creating sustainable fresh food supply chains through waste reduction. Int. J. Phy. Distrib. Logist. Manag. **43**, 262–276 (2013)
45. Rijpkema, W., et al.: Effective sourcing strategies for perishable product supply chains. Int. J. Phys. Distrib. Logist. Manag. **44**, 494–510 (2014)
46. Heising, J.K., et al.: Options for reducing food waste by quality-controlled logistics using intelligent packaging along the supply chain. Food Addit. Contam.: Part A **34**, 1672–1680 (2017)
47. Garrone, P., et al.: Reducing food waste in food manufacturing companies. J. Clean. Prod. **137**, 1076–1085 (2016)
48. Muriana, C.: A focus on the state of the art of food waste/losses issue and suggestions for future research. Waste Manag. **68**, 557–570 (2017)
49. Papargyropoulou, E., et al.: The food waste hierarchy as a framework for the management of food surplus and food waste. J. Clean. Prod. **76**, 106–115 (2014)

Decision Support Systems in Industrial and Business Applications

Time-Aware Knowledge Graphs
for Decision Making
in the Building Industry

Herwig Zeiner[1(✉)], Wolfgang Weiss[1], Roland Unterberger[1], Dietmar Maurer[1],
and Robert Jöbstl[2]

[1] JOANNEUM RESEARCH Forschungsgesellschaft mbH, Graz, Austria
herwig.zeiner@joanneum.at
[2] Haas Fertigbau GmbH, Großwilfersdorf, Austria
http://www.joanneum.at, http://www.Haas-Group.com

Abstract. The phenomenal progress in the development of the Internet
of Things (IoT) has already had tremendous impact on almost all industrial sectors and, finally, on our everyday life. The ongoing total digital
transformation leads to entirely new time-aware services and creates new
and more pro-active business opportunities.

Knowledge graphs are becoming more and more popular in different
domains, and also for the integration of sensor networks. In this project
we make knowledge graphs time-aware through a set of general temporal
properties relevant for the integration of sensing networks. Time-aware
knowledge graphs enable us to do time series analysis, find temporal
dependencies between events, and implement time-aware applications.
The requirements for the temporal properties derive from a use case of
residential real estate, with the aim to enable the occupants to interact
with their houses.

Keywords: Building industry · Time-aware analytics ·
Knowledge graph · Decision making

1 Introduction

The next generation of sensor networks analytics methods for many industries
will improve all steps within their business processes to be fast, versatile and
competitive. Connectivity alone, however, does not provide added value for applications and users. It is necessary to analyze the raw data to create meaningful
predictions and decisions in time. To make the right decisions on time, it is
vital to move the data processing away from centralized nodes into the edge of
networks, aiming at reducing the latency and minimizing essential information
exchange between sensors and the analytics components.

This paper focuses on the following aspects. First, we will develop a better time-aware support for applications. Thus, we need to implement a novel

© Springer Nature Switzerland AG 2019
P. S. A. Freitas et al. (Eds.): EmC-ICDSST 2019, LNBIP 348, pp. 57–69, 2019.
https://doi.org/10.1007/978-3-030-18819-1_5

time-aware semantics for all analytics components, enabling deterministic and therefore traceable execution of analytics workflows independent of their processing in the sensor devices. In addition, timing across interfaces will also require latency control among sensor devices and between crossing network domains.

Second, distributed decision making will allow users to make decisions faster and more accurate by providing meaningful knowledge models - presented by a time-aware knowledge graph - and coherent decisions for humans. Data will be collected and analyzed on various nodes in a sensor network and combined with context (e.g. time and space) information from knowledge graphs. The developed techniques also enable the exchange and replacement of the sensor nodes, including their knowledge, at runtime.

This paper is structured as follows. Section 2 explains background and related works. Afterwards, Sect. 3 describes our approach of building a knowledge graph. Section 4 explains the motivation for the chosen building scenarios. Afterwards in Sect. 5, we describe the use of the time-aware knowledge graph in the building domain. Finally, Sect. 6 presents the conclusion and the future plans for this work.

2   Background and Related Work

This section describes the state of the art in several relevant topics in the field of time-aware decision making for connected Internet of Things related systems in buildings [28]. The fields of research relevant for the mathematical foundation of planning and scheduling aspects of a multi-agent system involving teams of human and machine are mainly stemming from operation research including distributed optimization with graphs and also aspects of game theory [26].

2.1   General Time-Aware Aspects in Decision Making

Fluid and efficient decision making requires the handling of temporal aspects during the planning and the scheduling process. Recently, there has been an increasing interest in how to deal with temporal aspects in multi-agent decision setups. This aspect of temporal knowledge throughout is often expressed as a Temporal Network (TN) [17], where each activity (or task) is associated with two time points (e.g. start and end time). Temporal networks are a technique for modeling nodes and connections which appear and disappear at various timescales. Many of these models can be defined as connectivity driven, as the network's topology is at the core of the graph model [2]. One of the simplest networks can be described as Simple Temporal Network. Such Simple Temporal Networks (STNs) provide the starting point to deal with temporal constraints in planning and scheduling problems. STNs are typically mapped to the equivalent Distance Graphs (DGs) to check for the absence of negative cycles and thereby prove the controllability of the plan [9]. Recent work by Morris [23] considers back propagation rules to dynamically preserve the possibility for reasoning between interacting agents.

2.2   Distributed Systems and Time-Awareness

Monitoring and troubleshooting in distributed systems is a challenge. A key aspect in such networks consisting of sensor nodes, gateways nodes, computing nodes and, including, communication nodes is the time synchronization between the nodes [34]. To be able to merge event data from different types of nodes in a distributed system, it is necessary to introduce clocks and assign timestamps to the generated events. This requires all event producers to use the same time. Solutions could be to synchronize the individual clocks (e.g. by using the Network Time Protocol (NTP) [22]) of the event producers, or to have all event producers use the same clock (e.g. using the GPS time or a time server). The second approach is not always available, has some latency and might create varying time stamps. In general, it depends on the purpose and the temporal granularity at which events are produced to be able to decide if one of the above mentioned methods is feasible. (cf. [13, pp. 291–295], [10,24]).

One of the simplest ways to synchronize clocks is Christian's algorithm [7] which is basically a request/response method that depends on a single server and the request round-trip time. An improved approach is given in the Berkley Algorithm [14] in which clocks are synchronized by using several time sources. At this point, the Network Time Protocol (NTP) [22] is an evolution of the two previous methods. Unfortunately, it also may present some issues. It synchronizes the clocks in a predefined interval, but clocks may drift on their own, and the synchronization process has some inaccuracies too. The Precision Time Protocol (PTP) [1,11] is another way to synchronize a clock in a distributed system. This protocol is specified for elements connected to a local network or a few subnetworks in which high precision is required. The system is based on a master/slave methodology. Usually there are some master clocks which are able to be synchronized outside the current network, after which all other slave clocks are synchronized directly to these masters by using the request-response delays.

2.3   Time-Aware Aspects in Service Composition

Temporal aspects can also model in more complex business processes which are used in software intensive systems such as cloud computing. A general analysis on this time aspect can be found in the work of Lanz, which includes a state-of-the art analysis on time constraints in process-aware information systems [19]. Here, conflicts can occur when different services from different independent service processing chains compete for limited resources that have time constraints on services (e.g. minimal or maximal duration of a service) or the processing chain as a whole (e.g. minimal/maximal duration). It may also be important to know at what time the processing chain is ready without exceeding the maximal estimated time. Such complex jobs cannot be fulfilled by a single atomic service in a critical secure application environment; thus composite services built from interconnected jobs are required in many application domains [6,18,30]. Adaptive service composition and corresponding description languages describe how services can interact with each other, including the business logic and execution order of the interactions [37]. Automatic service composition is now an active area of research [21]. Several of the research aspects (e.g. workflow description, semantic description of the services, dynamic change of workflow, SLA management, monitoring and controlling [38]) in order to support SLA management are also relevant in the workflow optimization within the context of time-aware service composition [20].

2.4 Time-Aware Knowledge Graphs

Today, knowledge graphs are a backbone of many information systems that require access to structured knowledge. This knowledge can be domain-specific (e.g. building scenarios) or domain-independent. The idea of building intelligent systems with formalized knowledge of the world dates back to artificial intelligence research in the 1980s [25]. Knowledge graphs are a way representing complex relationships between real world entities. The concept received again attention (e.g. Linked Open Data [4], Google Knowledge graph, and Facebook's open graph [5]). Knowledge graphs can also be applied in the building sector [27].

Recently, the concept of knowledge graphs was enriched with time-aware knowledge for events and facts with duration for time-aware point of interest recommendations, as described in [36]. Time-awareness introduces a separate dimension to knowledge. Therefore, the temporal scope of facts is a relevant aspect in creating a knowledge graph. In the work of [33] the extraction of temporal facts from semi-structured data was described. Talukdar et. al. [31] proposed a method for discovering temporal ordering. In [8] a new method of inference including the prediction of temporal scopes for relational facts with missing time annotation was proposed.

Semantic representation of data is an important aspect not only in the development of knowledge graphs. The W3C Spatial Data on the Web Interest Group developed a recommendation for the Semantic Sensor Network Ontology [15]. This ontology already defines various properties for sensor data, but some more detailed temporal aspects are missing. Our approach complements this ontology with additional properties allowing in depth temporal analysis and processing of event streams.

2.5 Heuristics in Decision Making

Heuristic optimization methods are universally applicable optimization processes that determine high-quality solutions with realistic computational efforts and therefore contribute to the effective resolution of real multi-agent decision-making problems. In the first part, some heuristic priority rules are to be explained as well as common metaheuristics for the solution of Resource Constrained Project Scheduling Problems (RCPSP) [16] are to be considered. Since such methods have a faster resolvability than exact techniques, they are of great relevance, especially in practice. Common priority rules are, for example, the earliest due date (EDD) rule, used to minimize the delay, or Shortest Processing

Time (SPT), which tries to reduce the sum of the end times of all individual jobs. These rules are suitable for minimizing the total completion time. Heuristics, however, are also a common approach within multi-agent planning in general [12]. Recently, heuristic planning has made a lot of progress because of the use of powerful estimators. This progress, however, has not been translated to multi-agent planning due to the difficulty of applying classical heuristics in distributed environments. The application of local search heuristics in each agent has been the most widely adopted approach in multi-agent planning, but there have been some recent attempts to tackle the problem using global heuristics [32]. Another attempt was carried out in applying relaxation heuristics [29] for multi-agent planning.

3   Our Approach of Building a Knowledge Graph

Time-aware knowledge graphs are a prerequisite to realize Internet of Things applications which consider the concept within their applications. This type of applications are able to do temporal analysis of data. This can be (near) real-time analysis of data or time series analysis, or finding temporal dependencies between events - to name a view aspects. Things get more complex in a distributed system.

As an example Fig. 1 illustrates a generic architecture to demonstrate the use case introduced in Sect. 4.

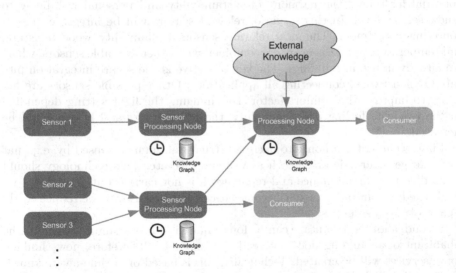

Fig. 1. A typical event-driven, distributed architecture with sensors, processing nodes and consumers. We use humidity, temperature, and CO_2 sensors. These sensors are placed all over the house. External sensors (e.g. weather station) can be integrated. Each processing node has its own knowledge graph and its own clock, which is synchronized to generate events if a certain value or time has been passed. This event will be forwarded to other nodes.

It is built as an event-based, distributed system, containing independent and locally distributed sensors and processing nodes. Sensors are the devices which analyze its environment, but sensors on its own cannot communicate with the rest of the system. This is done by sensor processing nodes, which process the raw sensor data and deliver enriched data to its consumers through the connected computer network (e.g. a Wireless LAN or any other available computer network). Data enrichment can be done as sensor processing nodes are intelligent devices, which store their state, have a local knowledge graph, and have a synchronized clock. This enables these devices to add the necessary time stamps to the resulting event data, and depending on the type of sensor, adding additional contextual or pre-processed information (e.g. average temperature over the last hour, temperature increase factor). Consumers of event data can be end user devices, or other processing nodes which combine information from other sensor processing nodes as well as from external data sources (e.g. a weather forecast). An important prerequisite for a working distributed system, as described here, is the synchronization of all nodes' clocks. There are various existing and well known approaches to do this. Methods for clock synchronization is described in Sect. 2.2 of this paper.

4   User Scenarios of the Building Industry

The goals of digitization, in our case for the timber building sector, offer great potential for taking the harmony of sustainability and personal well-being to a new service level. In this scenario, relevant sensors will be integrated into a house made of timber. The most relevant sensors are humidity, wood moisture, and temperature. These sensors are combined with other available sensors, which are already in use in buildings. The key objective is the secure integration into only ONE network, connecting all applications. Other possible sensors are lux meter to improve the comfort factor. By changing the light setting depending on the residents feeling or current duty, the life, comfort and health could be increased.

First, we monitor a house to decrease structural damage caused by, e.g., flat roof leakage, neuralgic areas such as window sills, etc. This technology should be used so that maintenance and repair work is not carried out according to a fixed schedule, but only when it becomes necessary. In this way, we reduce costly repairs of various defects.

Second, another advantage can be found in digital user manuals. Usually, the inhabitant of a building does not read user manuals. Therefore, new thinking house services will be created. Technically, this is based on a time-aware knowledge graph which helps the users to treat the buildings properly. The target is a self-learning open system which monitors, analyzes and then recommends the next best steps for their users. Also, if possible, connects several households and exchanges the knowledge on, how they kept their house and how to avoid certain failures. For example, there are continuously ventilating certain rooms to prevent mold formation. This might vary from house to house, depending on

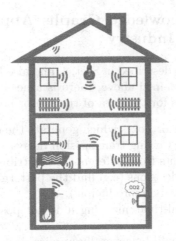

Fig. 2. Schematic real estate with distributed sensors of different kind like: temperature, lux meter, humidity;

location or usage, but hopefully a knowledge exchange will overall improve how the handling of mold and keep the cost down (Fig. 2).

Third, systematic misconstruction can be detected much faster and the training curve can be thus increased accordingly. This will reduce complaint costs for suppliers and customers, improve quality indicators for planning, purchasing, production and operation, and provide clearly defined maintenance and repair services.

The overall goal is to provide the house owner a system similar to cars. Giving them necessary advices how to handle the house or flat will reduce costs and troubles. Each sensor in the house is collecting data; each of them has a time stamp. The time- aware knowledge graph collects the data, saves them and analyzes these data. It can or should also be possible to integrate not only the house internal sensor data, but also additional external data. This could be, for example, weather data or, as mentioned above, data from other houses. Other houses could be of the same type or same neighborhood to enrich the data. Depending on the output of the analysis, the resident gets advices like, open the window now, to improve the ambient air, and/or avoid molt. Another example, could be the heating control, to optimize the room temperature depending on the residents daily life circle.

In a model house set up by Haas-Fertigbau in Austria, we implemented a hierarchical, extendable sensor-network based on inexpensive single-board computers running Linux for fast-prototyping and security reasons and equipped it with a range of low-cost, standard commercial digital sensors (e.g. DPS310 [digital air pressure], HDC1000 [humidity, temperature], CCS811 [VOC, CO2]) and analog sensors (e.g. distance meters). The network has been running and delivering data since August 2018.

5   Time-Aware Knowledge Graphs Applied to the Building Industry

Based on the user scenario described in Sect. 4, and also based on the distributed architecture in Sect. 3 explained above, there is a need to implement following time relevant requirements for the data of the sensor network:

- Nodes including the sensor boxes which generate the events are working on a computer network with varying transmission delays.
- Fuse this generated events from different and distributed nodes.
- Make temporally valid decisions (e.g. find the pattern "event A is followed by event B within 5 s" must be reproducible).
- Make decisions near real-time, meaning it is not possible to wait for a long time span.
- Temporal attributes should be generally applicable.

Now, we can define a set of general temporal event properties which are relevant for the data and the resulting higher level events.

- *Event occurrence time:* time when the event occurred in the real world.
- *Event detection time:* the time when the event producer detected that the event occurred.
- *Event emit time:* the time when the event leaves the event producer.
- *Event receive time:* the time when the event was received at the processing agent.

These properties are attached to each event which is generated by the processing nodes. Based on these temporal event properties, it is possible to calculate following durations:

- *Event detection duration:* is caused by the event detection mechanism, hardware, and algorithms (*event detection time - event occurrence time*).
- *Network transmission duration:* this delay is caused by the underlying network (*event receive time - event emit time*).
- *Full processing duration:* this is the duration when the event occurred in the real world until it was received in the processing node (*event receive time - event occurrence time*).

Optional, an additional property can be added to define the temporal validity of the event:

- *Valid until:* the expiration date of this knowledge artifact might be defined by the event producer.

Adding temporal aspects to knowledge graphs enables a range of new possibilities:

- Carry out temporal reasoning and apply Allen's temporal relationships [3] (e.g. event A followed by event B within 10 s).

- Identify outdated knowledge.
- Identify temporally revised knowledge.
- Any kind of time series analysis (e.g. identify trends).
- Bring event streams in correct temporal order (out-of-order event compensation [35]).
- Temporal traceability of resulting decisions (which events and facts where involved in the resulting decisions and when and in which order did they occur).
- Combine local knowledge with external knowledge which also might have temporal aspects: e.g. temperature in the house (local) and the weather forecast (external, was generated at a specific time point and has an expiration date).

5.1   Concept Graph of Temporal Event Properties

An concept graph (or ontology) was developed to implement the above defined temporal observations. We call it *Temporal Event Properties Graph (tep)* and it contains all time points which are relevant for events. The processing nodes add these properties to the events in the implement application. The temporal durations can be calculated on demand based on the time points and are not included in the graph. The following listing is an excerpt of the Temporal Event Properties Graph:

```
tep:occurrenceTime  a  owl:DatatypeProperty ;
rdfs:label  "event occurrene time"@en ;
rdfs:comment  "Time when the event occurred in
the real world."@en ;
rdfs:range  xsd:dateTime ;
rdfs:isDefinedBy  tep: .

tep:detectionTime  a  owl:DatatypeProperty ;
...

tep:emitTime  a  owl:DatatypeProperty ;
...

tep:receiveTime  a  owl:DatatypeProperty ;
...

tep:validUntilTime  a  owl:DatatypeProperty ;
...
```

5.2   Implementation of the Usage Examples

When processing event streams which are marked with the temporal event properties it is easily possible to implement a reusable temporal processing logic. In an experimental setup, we implemented components for the first three examples where we used these temporal properties. The other examples are highly relevant and will be further investigated in the future.

(1) An out-of-order event compensation which re-orders event streams in a processing node based on the occurrence time, to make sure that a subsequent temporal processing (e.g. event processing, time series analysis) can be done correctly. Therefore, we used a buffer algorithm which dynamically adapted its buffer size (see also [35]) according to the full processing duration.

(2) A notification event for the house user was generated when the following temporal condition was met: window open event followed by air-quality good event within 15 min. This tells the user that the window can be closed as the air quality is good again within the room. If the 15 min are exceeded, the user will get a notification anyway.

(3) Analyzing temperature data especially on the house internally, but also on weather data, allows us to identify unusual outliers in the time series. There can be different types of outliers. One being that the temperature does not change at all over a period of time, so we consider the sensor is malfunctioning. If the temperature changes and is more out of range (lower or higher than usual) we state that there must be an anomaly. That can be an open window or the overheating of the room. If there is similar data throughout the house, the challenge is to know if it is just an unusual day or if there is another reason (e.g. resident is on vacation and the overall room temperature is set lower).

(4) By using different sensor types like room temperature and humidity, ambient temperature and humidity as well as external data like weather forecast and fuse these sensor values together. The goal is to create an event driven heating and ventilation to improve the overall house comfort level. Even historical or external data from nearby households should improve the timing, when the windows should be opened and how long they should stay open. Sensors capable of sensing CO_2 or other fine dust elements will also be included in the event driven fresh air circulation. On the other side, it should also reduce possible heating cost during cold days by opening the windows when the occupant is leaving his home or does not use the specific room.

6   Conclusion and Future Work

In this paper we discussed a time-aware knowledge graph for decision making in the envisioned building sector. We have described the scenarios in the building sector in detail. Afterwards, we defined a set of time-relevant requirements for the time-aware knowledge graph. These requirements are the ingredients for the Temporal Event Properties Ontology. With this ontology we build the knowledge graph which can be used in various system components (e.g. processing nodes).

In our future work, the implementation of this time-aware knowledge graph into the system will be extended to embrace new scenarios in which more complex time patterns are needed. We intend to develop what we consider to be a toolbox for time-aware distributed decision by using a time-aware knowledge graph from which we can choose the one that is the most appropriate for a given building scenario.

Acknowledgement. The K-Project Dependable, secure and time-aware sensor networks (DeSSnet) is funded within the context of COMET – Competence Centers for Excellent Technologies by the Austrian Ministry for Transport, Innovation and Technology (BMVIT), the Federal Ministry for Digital and Economic Affairs (BMDW), and the federal states of Styria and Carinthia. The program is conducted by the Austrian Research Promotion Agency (FFG). The authors are grateful to the institutions funding the DeSSnet project and wish to thank all project partners for their contributions.

References

1. IEEE Standard for a Precision Clock Synchronization Protocol for Networked Measurement and Control Systems. IEEE Std 1588–2008, pp. 1–269, July 2008. https://doi.org/10.1109/IEEESTD.2008.4579760
2. Albert, R., Jeong, H., Barabási, A.L.: Internet: diameter of the world-wide web. Nature **401**(6749), 130 (1999)
3. Allen, J.F.: Maintaining knowledge about temporal intervals. Commun. ACM **26**(11), 832–843 (1983). https://doi.org/10.1145/182.358434
4. Auer, S., Bizer, C., Kobilarov, G., Lehmann, J., Cyganiak, R., Ives, Z.: DBpedia: a nucleus for a web of open data. In: Aberer, K., et al. (eds.) ASWC/ISWC -2007. LNCS, vol. 4825, pp. 722–735. Springer, Heidelberg (2007). https://doi.org/10.1007/978-3-540-76298-0_52
5. Bordes, A., Gabrilovich, E.: Constructing and mining web-scale knowledge graphs: KDD 2014 tutorial. In: Proceedings of the 20th ACM SIGKDD International Conference on Knowledge Discovery and Data Mining, p. 1967. ACM (2014)
6. Chen, N., Cardozo, N., Clarke, S.: Goal-driven service composition in mobile and pervasive computing. IEEE Trans. Serv. Comput. **11**(1), 49–62 (2018)
7. Cristian, F.: Probabilistic clock synchronization. Distrib. Comput. **3**(3), 146–158 (1989). https://doi.org/10.1007/BF01784024
8. Dasgupta, S.S., Ray, S.N., Talukdar, P.: HyTE: hyperplane-based temporally aware knowledge graph embedding. In: Proceedings of the 2018 Conference on Empirical Methods in Natural Language Processing, pp. 2001–2011 (2018)
9. Dechter, R., Meiri, I., Pearl, J.: Temporal constraint networks. Artif. intell. **49**(1–3), 61–95 (1991)
10. Della Valle, E., Schlobach, S., Krötzsch, M., Bozzon, A., Ceri, S., Horrocks, I.: Order matters! Harnessing a world of orderings for reasoning over massive data. Semant. Web **4**(2), 219–231 (2013). http://dl.acm.org/citation.cfm?id=2590215.2590219
11. Eidson, J.C.: Measurement, Control, and Communication Using IEEE 1588, 1st edn. Springer, London (2006). https://doi.org/10.1007/1-84628-251-9
12. Ephrati, E., Rosenschein, J.S.: A heuristic technique for multi-agent planning. Ann. Math. Artif. Intell. **20**(1–4), 13–67 (1997)
13. Etzion, O., Niblett, P.: Event Processing in Action, 1st edn. Manning Publications Co., Greenwich (2011)
14. Gusella, R., Zatti, S.: The accuracy of the clock synchronization achieved by TEMPO in Berkeley UNIX 4.3BSD. IEEE Trans. Softw. Eng. **15**(7), 847–853 (1989). https://doi.org/10.1109/32.29484
15. Haller, A., Janowicz, K., Cox, S., Phuoc, D.L., Taylor, K., Lefrançois, M.: Semantic sensor network ontology. W3C recommendation, W3C, October 2017. https://www.w3.org/TR/2017/REC-vocab-ssn-20171019/

16. Herroelen, W., De Reyck, B., Demeulemeester, E.: Resource-constrained project scheduling: a survey of recent developments. Comput. Oper. Res. **25**(4), 279–302 (1998)
17. Holme, P., Saramäki, J.: Temporal networks. Phys. Rep. **519**(3), 97–125 (2012)
18. Lahmar, F., Mezni, H.: Multicloud service composition: a survey of current approaches and issues. J. Softw.: Evol. Process **30**(10), e1947 (2018)
19. Lanz, A., Weber, B., Reichert, M.: Time patterns for process-aware information systems. Requirements Eng. **19**(2), 113–141 (2014)
20. Ben Mabrouk, N., Beauche, S., Kuznetsova, E., Georgantas, N., Issarny, V.: QoS-aware service composition in dynamic service oriented environments. In: Bacon, J.M., Cooper, B.F. (eds.) Middleware 2009. LNCS, vol. 5896, pp. 123–142. Springer, Heidelberg (2009). https://doi.org/10.1007/978-3-642-10445-9_7
21. Van der Mei, R., et al.: State of the art and research challenges in the area of autonomous control for a reliable internet of services. In: Ganchev, I., van der Mei, R.D., van den Berg, H. (eds.) Autonomous Control for a Reliable Internet of Services. LNCS, vol. 10768, pp. 1–22. Springer, Cham (2018). https://doi.org/10.1007/978-3-319-90415-3_1
22. Mills, D.L.: Computer Network Time Synchronization: The Network Time Protocol on Earth and in Space, 2nd edn. CRC Press Inc., Boca Raton (2010)
23. Morris, P.: Dynamic controllability and dispatchability relationships. In: Simonis, H. (ed.) CPAIOR 2014. LNCS, vol. 8451, pp. 464–479. Springer, Cham (2014). https://doi.org/10.1007/978-3-319-07046-9_33
24. Neville-Neil, G.V.: Time is an illusion lunchtime doubly so. Commun. ACM **59**(1), 50–55 (2015). https://doi.org/10.1145/2814336
25. Russell, S.J., Norvig, P.: Artificial Intelligence: A Modern Approach. Pearson Education Limited, Malaysia (2016)
26. Shoham, Y., Leyton-Brown, K.: Multiagent Systems: Algorithmic, Game-Theoretic, and Logical Foundations. Cambridge University Press, Cambridge (2008)
27. Soliman, M., Abiodun, T., Hamouda, T., Zhou, J., Lung, C.H.: Smart home: integrating internet of things with web services and cloud computing. In: 2013 IEEE 5th International Conference on Cloud Computing Technology and Science (CloudCom), pp. 317–320. IEEE (2013)
28. Stojkoska, B.L.R., Trivodaliev, K.V.: A review of internet of things for smart home: challenges and solutions. J. Cleaner Prod. **140**, 1454–1464 (2017)
29. Stolba, M., Komenda, A.: Relaxation heuristics for multiagent planning. In: ICAPS (2014)
30. Strunk, A.: QoS-aware service composition: a survey. In: 2010 Eighth IEEE European Conference on Web Services, pp. 67–74. IEEE (2010)
31. Talukdar, P.P., Wijaya, D., Mitchell, T.: Acquiring temporal constraints between relations. In: Proceedings of the 21st ACM International Conference on Information and Knowledge Management, pp. 992–1001. ACM (2012)
32. Torreno, A., Sapena, O., Onaindia, E.: Global heuristics for distributed cooperative multi-agent planning. In: ICAPS, pp. 225–233 (2015)
33. Wang, Y., Zhu, M., Qu, L., Spaniol, M., Weikum, G.: Timely YAGO: harvesting, querying, and visualizing temporal knowledge from Wikipedia. In: Proceedings of the 13th International Conference on Extending Database Technology, pp. 697–700. ACM (2010)
34. Weiss, M.A., et al.: Time-aware applications, computers, and communication systems (TAACCS). Technical report (2015)

35. Weiss, W., Jiménez, V.J.E., Zeiner, H.: A dataset and a comparison of out-of-order event compensation algorithms. In: IoTBDS, pp. 36–46 (2017)
36. Yuan, Q., Cong, G., Ma, Z., Sun, A., Thalmann, N.M.: Time-aware point-of-interest recommendation. In: Proceedings of the 36th International ACM SIGIR Conference on Research and Development in Information Retrieval, pp. 363–372. ACM (2013)
37. Zeiner, H., Halb, W., Lernbeiß, H., Jandl, B., Derler, C.: Making business processes adaptive through semantically enhanced workflow descriptions. In: Proceedings of the 6th International Conference on Semantic Systems, p. 27. ACM (2010)
38. Zhang, P., Jin, H., He, Z., Leung, H., Song, W., Jiang, Y.: IgS-wBSRM: a time-aware web service QoS monitoring approach in dynamic environments. Inf. Softw. Technol. 96, 14–26 (2018)

A Conceptual Business Process Entity with Lifecycle and Compliance Alignment

George Tsakalidis[1] , Kostas Vergidis[1]([⊠]) , Pavlos Delias[2] ,
and Maro Vlachopoulou[1]

[1] Department of Applied Informatics, University of Macedonia,
Thessaloniki, Greece
{giorgos.tsakalidis,kvergidis,mavla}@uom.edu.gr
[2] Department of Accounting and Finance,
Eastern Macedonia and Thrace Institute of Technology, Kavala, Greece
pdelias@teiemt.gr

Abstract. This paper proposes a conceptual model that incorporates: (a) a business process entity, (b) a business process lifecycle aligned with the proposed entity, and (c) a compliance framework that focuses on the degree of process compliance to imposed regulatory standards. The pursuing objective is to systematize business processes through a conceptual entity applicable to Business Process Management (BPM) practices and compliance-checking. The applied methodology involves the review, interpretation and comparison of business process definitions, structural elements and their interrelations, acknowledged BPM lifecycles and compliance rules. The initial findings lead to the proposal of a contextual business process structure that sets the boundaries of business process as a clearly defined entity. The business process entity encompasses continuous modification of its design based on the feedback it generates. Additionally, a comparative analysis of prominent BPM lifecycles resulted in a proposed 'business process lifecycle' that allows for a better alignment of the included cycle steps. The proposed business process entity is also related with process compliance practices to produce compliance-aware business processes. The introduced conceptual model can assist professionals in apprehending core business process features, focusing on process flexibility and redesign. It can also serve as a preliminary prototype for checking the degree of compliance between a business process and the applicable compliance rules.

Keywords: Business process · Business Process Management ·
BPM lifecycle · Business process lifecycle · Business process compliance

1 Introduction

Business processes are conceived as a set of related activities, orderly performed for actualizing a business objective. Through maintaining an orientation towards business processes, organizations orchestrate and achieve continuous improvements of their activities on time and within specified resource constraints, in an effort to gain competitive superiority [1]. One of the fields dealing with the induced challenges is BPM that has emerged as an area of high interest among scholars, professionals and

© Springer Nature Switzerland AG 2019
P. S. A. Freitas et al. (Eds.): EmC-ICDSST 2019, LNBIP 348, pp. 70–82, 2019.
https://doi.org/10.1007/978-3-030-18819-1_6

practitioners [2]. BPM typically consists of a sequence of discrete activities for the continual improvement of business processes, carried out within an iterative life cycle [3]. The continuous research on BPM resulted in a plethora of methods, techniques, and tools to support the design, enactment, management, and analysis of operational business processes [4]. Although the practical relevance of BPM has proved undisputed, debates still accrue and persist, regarding the identity, quality and maturity of the BPM field [5].

The authors aim to address the ambiguity of business process as a concept and the resulting approaches in BPM lifecycles. This is achieved by proposing a contextual business process structure that ratifies business process as a unique entity, that encompasses continuous modification and evolution of its design, based on the feedback it generates. The authors also propose a lifecycle model aligned to the business process entity. The lifecycle should be attributed to a business process and the various stages it evolves through; not BPM as an entity. Treating the lifecycle as a process itself will reveal a closer and more detailed interaction of the various cycle steps and will provide a clearer perspective of how a business process progresses and what tools and technologies are better suited for each of its stages. Lastly, compliance-checking in the proposed lifecycle can generate business processes applying a novel 'compliance-by-design and redesign' approach. The conceptual model discussed in this paper embodies three main notions:

- a *business process entity* that sets the boundaries of a business process as a clearly defined entity. The proposed structure portrays the business process entity as continuously evolving and adapting its design based on the execution feedback that it generates. The specific approach towards design modification (whether it is redesign, improvement, restructuring, or formal optimization) is left open to different disciplines and methodologies.
- a *business process lifecycle* aligned to the business process structure that emphasizes the value of adaptability. The various cycle steps are extracted from critically reviewing the existing BPM lifecycles and they are unified in the proposed structured to better highlight that a clearly defined concept of business process also reflects the stages that it is flowing through.
- a *compliance-aware framework* that addresses the need for organizational business processes to comply to various internally and externally imposed regulatory frameworks. The authors reviewed existing approaches and propose a combination of a priori and a posteriori compliance approaches and also highlight where these connect to the proposed business process entity.

2 Related Work

The increased popularity of business process has resulted in a variety of interdisciplinary approaches [6], a fact that also underpins the ambivalent nature of the business process scope. Völkner and Werners [7] believe that there is no generally accepted business process definition due to the fact that the concept has been engaged by a number of different disciplines. Moreover, Lindsay et al. [8] underline that business process definitions are based on machine metaphor type explorations of a process, suggesting that most of them are limited in depth, leading to constrained corresponding models.

Authors such as Van Der Aalst et al. [4] and Dumas et al. [9], attempt to rationalize the ambiguity in generic business process definitions by deploying a set of components that structure a business process. The purpose of a business process is the processing of various cases (e.g. online orders, sales and calculation of travel expenses) that can be either too simplistic when restricted to a functional unit of an organization or more complex by cutting across several business partners [10]. According to Van der Aalst [11] there are two important elements for a business process to be defined: (a) the activities, that are usually a set of tasks in a specific order, and (b) the allocation of resources to these tasks. Similarly, Dumas et al. [9] indicate that a business process encompasses a number of events and activities, through illustrating a typical business process example. Other perceived components are: (i) process structure (i.e. control flow, data flow dependencies and business rules that cover execution constraints), (ii) process goals, and (iii) structural elements such as resources, input and output [12]. The combination of these components and their relationships construct a structure that attempts to formalize a business process and transfuse a much-desired uniqueness in terms of operations perspective. However, most of these approaches are not extensive on the components they employ [11], they result in either too simplistic [13] or too complex structures ([14], [15]) and undermine the capability for effectively redesigning a business process as they capture mostly static elements.

The comprehension of the different tools, techniques, terminologies and features of BPM allowed for the conceptualization of what is referred to as BPM lifecycle [16]. Business Process Management (BPM) encompasses a set of methods, techniques, and tools for handling business processes (i.e. modeling, execution and analysis) of an organization [3], which are organized in phases and steps, referred to as BPM lifecycle [9]. Advocates of the BPM lifecycles propagate schematic diagrams that systematize the methodology and steps of a BPM project, in an effort to manage effectively the organizational operations. BPM lifecycles are continuous [17] and composed of activities [18]. However, there are multiple variations and convergences throughout literature [19] regarding of what is actually included in such lifecycle. Researchers either propose simple sequential diagrams with rigid connections between the different lifecycle steps, or introduce illustrations with multi-faceted interfaces in an effort to achieve specific objectives. A comparative survey of the established BPM lifecycles [20] demonstrates the abundance of existing approaches. Further investigation of these lifecycle models reveals a unique orientation for each e.g., how to incorporate external factors into BPM [21], or how to analyze whether BPM systems actually support the different phases of the lifecycle [22]. This variety of approaches underlines the absence of a unanimous point of view in the academic and business community, which results in limited and fragmented benefits.

Moreover, organizations face many different sets of regulations they have to comply with, from high-level regulations and frameworks (e.g., CMMI, ITIL, COBIT and ISO rules) to low-level specific business rules. An instance of non-compliance can result in severe penalties such as financial or reputation loss stressing the importance of the semantically correct and error-free execution of business processes. The proper executability of a process model has been based on semantic constraints [23]. These Semantic Constraints or Compliance Rules stem from standards, regulations, guidelines, corporate policies and laws and reflect whether a process complies with them.

Over the last decades, the global discipline of BPM academics, practitioners and professionals have incrementally improved established BPM lifecycle models. For BPM to remain relevant, the community has to reconsider its overall aim and make a shift from the economy of corporations to a digitally empowered economy of people [24]. This entails that the current perception of business processes, i.e. rigid rule-based execution of pre-defined activities has to evolve into dynamically evolving processes based on real-time customer needs and comprised of context-sensitive activities. This paper intends to highlight the need for a conceptual model tailored to current and emerging requirements. This context-aware model focuses on constant adaptation and applies for both business process lifecycle and compliance checking.

3   Business Process as an Entity: A Conceptual Model

Given the diversity of both theoretical (definitions, design) and functional (lifecycle, compliance) aspects of BPM, the authors propose a contextual business process structure (Fig. 1) that perceives business process as a unique entity and supports continuous modification and evolution of its design, based on the feedback it generates. The inspiration for the proposed approach is based on the description introduced in [8]: *"Sustainable business processes carried out by human operators are a balancing act between learning from the past and experimenting with and adapting to the future, and between rules and constraints versus freedom and flexibility"*. The proposed entity is separated in three distinct sets of components: (i) the prerequisite components, (ii) the process and contextual components, and (iii) the goal components.

We argue that for designing a business process, the prerequisite components are essential in determining the scope and the outcome: (a) who is the recipient of the outcome? The customer (external or internal to the organization) is the initiator of a particular business process instance and -in most cases- the recipient of its outcome. (b) What is the expected outcome of the process (i.e. product, service or a combination)? The desired outcome(s) of the process should be explicitly documented including the cases that they fail to be produced. Designing a business process without clear indication of what it produces or when it concludes is a recipe for disaster [25]. (c) What are the required resources and conditions for the outcome to be produced and the customer to be satisfied? This prerequisite is about specifying the necessary resources and conditions for the process to run smoothly and document the effect of their absence to a process instance. Completing the specification of the first set of components, the process designer has a clear idea on the business process: who is intended for, what is the outcome and what is required for its enactment.

The second set of components includes the construction of the process design along with the contextual elements. Examples of how context is incorporated in the business process specification include frameworks that describe context factors with relevance to BPM projects based on their settings [26], and select methods and mechanisms to work together for supporting context-aware BPM [27]. Relevant research is still at an early stage which provides the opportunity of formally incorporating context awareness into BPM. The other component is business process design that includes: (i) the business objectives (qualitative or quantitative) over which the process is evaluated; (ii) the particular arrangement of its structural elements (e.g., activities and other artifacts); and (iii) the data flows, i.e. the circulation of data throughout the process execution [28].

The business process instance is a specific enactment and implementation of the business process design, depending on the particular process inputs [29]. During a single execution, specific decisions are taken, following the actual events that lead to the performance of explicit activities. The combination of these elements produces each time a unique blueprint based on the initial design. It is important to acknowledge the static nature of a business process design, no matter how accurately describes an operation, and the dynamic knowledge-intensive nature of its generated instances that can provide a better understanding of the flexibility that a business process can demonstrate.

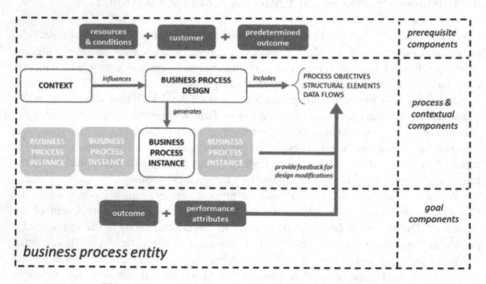

Fig. 1. Proposed structure of the business process entity

The goal components include: (i) the final outcome; complete or incomplete in relation to the predetermined one, and (ii) the performance attributes; multiple factors and key performance indicators, such as process efficiency, effectiveness and flexibility, policy adherence and traceability [30]. These attributes assess each process instance and provide feedback for the process performance. Based on the proposed entity, the process design is continuously modified based on the feedback of the goal components. Regarding design modification, we avoided using the words 'improvement, 'optimization' and 'redesign' as they require specific criteria, objectives and techniques. The aim of the proposed structure is to portray that the business process entity should encompass a continuous modification and evolution of its design based on the feedback it generates. The specific approach towards design modification (whether it is redesign, improvement, restructuring, optimization) is open to different disciplines and methodologies. The main novelties of the proposed structure of the business process entity are: (1) The business process entity has clear boundaries and interrelated components; (2) Context awareness in the design adds flexibility and highlights the subjective nature of each instance; (3) A predetermined outcome, specific performance attributes and the potential of design modification enable the capability for continuous process improvement and the application of various approaches towards modifying/improving the business process. The main rationale is that metadata, new insights and any kind of feedback generated during past executions can be leveraged to enhance the design of processes.

4   A Business Process Lifecycle Based on the Proposed Entity

The notion of BPM and the abundance of proposed lifecycles create ambiguity on structuring and managing an organization's business processes [31]. The authors propose that lifecycle should be attributed to a business process and the various stages (e.g., elicitation, modelling, enactment, redesign) it evolves through; not BPM as an entity. It is far easier to conceive and manage such perspective: an organization designs business processes by explicitly specifying the lifecycle, span and stages they evolve by utilizing the appropriate tools and methods in each phase. Many of the existing BPM lifecycles discussed in [20] make more sense under this perspective. Based on the most commonly used lifecycles in BPM, the authors suggest a comprehensive set of steps that compose the *business process lifecycle*:

1. *Specification:* This first step encompasses the specification of scope, i.e. the pre-determined outcome the customer intends for, along with the identification of explicit conditions and resources needed to be in place for a continual process execution. This step also serves as a primary conformance check that inspects the organizational capacity required for the next step.
2. *Design and modelling:* This step determines the artifacts, business objectives and data flows. Design refers to aligning the scope of the process to specific business operations, departments and tasks, whereas modelling refers to capturing the process using structured techniques with formal syntax. At this step, the particular organizational goals are specifically determined.
3. *Contextualization* (or configuration): A step that involves system selection and testing through selecting the subjective elements that influence the business process in a particular context. This phase takes place before the actual business process implementation and once it is completed, the system is launched in its context where the design is fine-tuned in accordance to the environment that the process will be enacted [3].
4. *Implementation, Execution and Monitoring:* The selected design is translated to an actual workflow taking place in the organization with the assistance of a Business Process Management System (BPMS). A process is enacted each time in the form of a unique process instance containing additional run-time information that can

provide feedback for the evaluation of performance. This step also encompasses the capability of switching variants during runtime to adapt to context changes [32].

5. *Performance analysis and evaluation:* The execution data collected from the various process instances in the previous step, are collected in log file and evaluated based on specific criteria that can be qualitative or quantitative. The competence of the execution environment is also analyzed for providing an assessment of contextual features.

6. *Redesign:* This step occurs through the application of various techniques and approaches that result in modifying the process design based on the feedback for the process run-time and/or the performance attributes e.g., to adapt to current conditions or optimize according to a given objective function. This stage can result in both high- and low-level design modifications, or complete overhaul of the business process depending on the technique utilized.

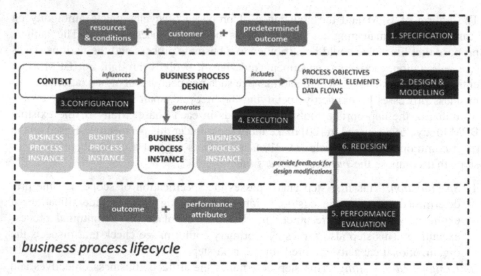

Fig. 2. Proposed business process lifecycle projected on the business process entity

The selection of steps for the proposed business process lifecycle comprises an inclusive collection in comparison to the existing lifecycles. It should be highlighted that the steps selected are first class citizens in all models presented in [20], despite the fact that they are incorporated into these models in different ways. Figure 2 extends the proposed business process entity by projecting the proposed lifecycle steps with the business process elements identified in the previous section. This further showcases that a concise definition and structure of what consists a business process encompasses the various cycle steps it evolves through its lifecycle alleviating the need to define processes and BPM lifecycles separately. The business process lifecycle adds on the proposed conceptual model by aligning its encompassing steps on the proposed entity thus bridging the gap between business process blueprints and existing BPM lifecycle models.

5 Multi-stage Compliance Framework

The final view of the conceptual model focuses on the topic of business process compliance. Governatori and Rotolo [33] define Business Process Compliance (BPC) as the relationship between two sets of specifications: (a) the specification for the processes adopted by a business to achieve its goal (i.e. the formal representation of a process model), and, (b) the specifications corresponding to the regulations relevant for a business (i.e. the formal representation of a relevant regulation). The authors propose an aligned approach on how compliance rules can be incorporated in the different components of the business process entity.

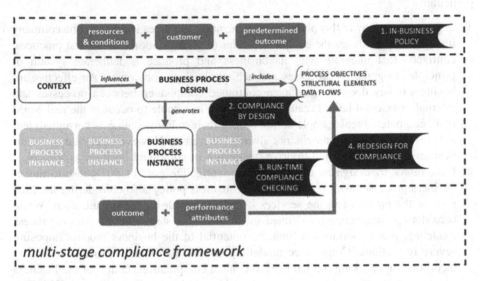

Fig. 3. Proposed compliance framework based on the business process entity.

According to Ghanavati et al. [33], compliance in business process entails:

1. Identification of the compliance rules that are relevant to a particular business process,
2. Tracking of compliance rules relevant to the organization due to the dynamically changing of legislatures and compliance requirements,
3. Extracting relevant legal requirements from source documents (i.e., legal texts),
4. Identification of noncompliance instances and prioritization for resolution to prevent the occurrence of non-compliant situations,
5. Modification of the process instance behavior the during its execution,
6. Integration of relevant legal requirements to the business processes of the organization and optimization of the overall business strategy,
7. Updating all of the above in the face of ongoing evolution of regulations and the organization itself.

Much of the work discussed in relevant literature attempts to ensure compliance between business processes and regulations [35]. The majority of the compliance-checking approaches focus on a particular checking moment [23, 36] during the business process lifecycle:

- a priori, including Forward Compliance Checking (FCC) and Design-Time Compliance Checking (DTCC), and,
- a posteriori, including Backward Compliance Checking (BCC).

Based on the above, Fig. 3 extends the proposed business process structure through relating business process compliance practices with the business process elements identified in the previous section to produce compliance-aware business processes. In particular:

1. *In-business policy:* in this phase, the organization produces its policy. The company identifies and analyzes the compliance rules (e.g., regulations, laws, best practices, contracts) that integrate with its business and produce a deliberate system of principles to guide its processes and achieve business outcomes. An effective in-business policy does not guarantee continued compliance between processes and internal or external laws, because problems are inevitable to occur in the real world (e.g., computers break, people do not absorb their training, data gets corrupted) and as a result a business policy is not always adhered to. However, generating the in-business policy is essential to drive the design of compliant business processes.

2. *Compliance by design or Compliance modeling:* the goal of this phase to model business processes with respect to the in-business policy also considering the final goal of the process and the services it has to provide to the organization. While considering compliance (as defined in the previous step) in the business process modeling, it is important not limit the potential of the business process imposing severe restrictions. Compliance modelling should also benefit process design by incorporating checking mechanisms to control compliant execution of the business processes. The aim of this stage is to achieve compliance by design [37]. The outcome of this compliance stage is a structured collection of mechanisms and other parameters that encompass business rules checked at run time.

3. *Run-time compliance checking:* this phase starts when the compliant processes are executed using a configured system (e.g., a WFMS). Run-Time Compliance Checking (RTCC) techniques such as Compliance alerts and Recovery actions ensure compliance of business processes with regulations and business rules on time to avoid ending in a non-compliant result. Such checking mechanisms find a non-compliant factor while running the process and allow process engineers to both control and monitor compliance rules during the generation of process instances [36]. This phase also calculates the degree of compliance of an executed compliant business process instance. During process execution, relevant data are collected and analyzed to determine how well the process is complying with respect to in-business policy and other compliance parameters. Namely, the data logged by the information system are used to diagnose the operational process bottlenecks and recurrent errors or deviations. Consequently, corrective actions are can be undertaken.

4. *Process Redesign for Compliance:* the goal of this phase is to monitor the aforementioned changes, analyze and apply them to a process model with increased in-
business policy compliance and alignment with the organizational performance
objectives. Redesign for compliance aims at a redesigned process with increased
degree of compliance which serves as a basis for the next iteration.

Also, it is important to highlight the need for an external consultant (e.g., a financial
auditor) to certify business compliance. This external actor physically visits the company and checks (i) whether the company has correctly interpreted the existing leg-
islation, (ii) whether business processes have been correctly implemented, and, finally,
(iii) whether they are correctly executed [38]. The system must be prepared to allow the
execution of internal audits aimed at performing routine controls by the organization,
and external audits carried out by compliance experts unconnected with the organization and responsible for checking whether the organization complies with the rules
[35].

The proposed compliance conceptual framework provides full-coverage of the
business process lifecycle to fulfill the need for automated compliance support. This is
achieved by combining before-the-fact, run time and after-the-fact compliance
checking approaches. More particular, in-business policy is a preventative step to avoid
non-compliant situations at the modeling phase. It's a mechanism (i.e., a collection of
legal norms relevant to our organization or originating from the government, stan-
dardization bodies, customers and the like) guaranteeing that a future process design
will be compliant. Compliance by design or Compliance modeling is a subsequent
verification of compliance while modeling the business process to result in a business
process that is compliant by design. In the following phase (i.e., run-time compliance
checking) the execution of business process (i.e., when process and event definitions
are consumed) is covered by checking compliance step by step (after each task is
executed). Finally, the compliance-aware framework goes a step forward by proposing
the stage of Process Redesign for Compliance which provides a means of evaluating
the impact of aforementioned compliance control phases, detect and analyze compliance violations on executed process models, and, provide feedback for process (re)
design.

6 Discussion and Conclusions

The authors put forward a three-fold conceptual model to: (i) clearly define the
boundaries and components of business processes, (ii) align and better manage its
lifecycle by matching specific cycle steps to corresponding elements, and (iii) generate
business processes applying a novel 'compliance-by-design and redesign' approach.
The proposal addresses current limitations and is amenable to further extensions. More
specifically, the authors examined core aspects of business processes; their definitions
and structure, the elements of a business process lifecycle and their interrelations. By
examining existing definitions and structures of business processes, the authors proposed an entity that is more contemporary in the sense that includes contextual

elements and focuses on continuous and adaptive process modification. A next step would be to develop a mechanism of recording and generating business process paradigms based on the proposed structure. In addition, a known issue in evaluating qualitative and quantitative approaches is the lack of a library of comparable business process problems. An established library of theoretical problems is common in many disciplines in testing the performance and consistency of new algorithms and techniques. Having a mechanism to capture and generate business process paradigms will assist in better evaluating the various modelling and improvement approaches put forward by researchers and practitioners in providing comparable results. Currently, the authors investigate business process paradigms used in literature as motivational examples, to compose a cumulative library that can initially serve to test and validate the conceptual framework. Thereafter, the proposed business process structure can be a starting point for creating test cases of business process designs that can then be utilized in specific domains.

This paper also examined the necessity for managing the lifecycle of business processes in a structured and coherent way. The authors proposed a lifecycle centered on the proposed business process entity. Most of the current approaches depict the cycle steps in a simple sequential manner failing to properly depict their complex interrelations. Our proposal matches the various lifecycle steps to specific components of the business process entity and aims for a more comprehensive approach in managing the process lifecycle. As a future direction, this approach will be further extended by modeling the interactions of the various steps through the lifecycle of a business process. Treating the BPM lifecycle as a process itself can potentially provide: (i) a closer and more detailed interaction of the various cycle steps, (ii) a clearer perspective of how a business process unfolds and (iii) the appropriate tools and techniques that are better suited for each stage.

Furthermore, the authors are currently focusing their efforts towards automated (or semi-automated) checking of the degree of compliance between a business process and the applicable compliance rules. The areas of focus, according to [34, 35]: (i) identification of particular protocols on compliance analysis (e.g., how to extract legal requirements, how to map them to business processes); (ii) construction of templates and samples to help organizations build compliant processes more easily; and (iii) formal extraction of compliance rules (i.e., in-business standard policy). A future extension of the proposed multi-stage compliance framework should address the above challenges.

References

1. Tsakalidis, G., Vergidis, K.: Towards a comprehensive business process optimization framework. In: 2017 IEEE 19th Conference on Business Informatics (CBI), pp. 129–134. IEEE (2017)
2. Hassani, A., Ghannouchi, S.A.: Analysis of massive e-learning processes: an approach based on big association rules mining. In: Park, J.H., Shen, H., Sung, Y., Tian, H. (eds.) PDCAT 2018. CCIS, vol. 931, pp. 188–199. Springer, Singapore (2019). https://doi.org/10.1007/978-981-13-5907-1_20

3. Weske, M.: Business Process Management: Concepts, Languages, Architectures. Springer, Heidelberg (2012). https://doi.org/10.1007/978-3-642-28616-2
4. Van Der Aalst, W.M., Ter Hofstede, A.H., Weske, M.: Business process management: a comprehensive survey. ISRN Softw. Eng. **2013**, 20 (2013)
5. Recker, J., Mendling, J.: The state of the art of business process management research as published in the BPM conference. Bus. Inf. Syst. Eng. **58**, 55–72 (2016)
6. Georgoulakos, K., Vergidis, K., Tsakalidis, G., Samaras, N.: Evolutionary multi-objective optimization of business process designs with pre-processing. In: 2017 IEEE Congress on Evolutionary Computation (CEC), pp. 897–904. IEEE (2017)
7. Völkner, P., Werners, B.: A decision support system for business process planning. Eur. J. Oper. Res. **125**, 633–647 (2000)
8. Lindsay, A., Downs, D., Lunn, K.: Business processes—attempts to find a definition. Inf. Softw. Technol. **45**, 1015–1019 (2003)
9. Dumas, M., Rosa, M.L., Mendling, J., Reijers, H.A.: Fundamentals of Business Process Management. Springer, Heidelberg (2018). https://doi.org/10.1007/978-3-662-56509-4
10. Pourshahid, A., et al.: Business process management with the user requirements notation. Electron. Commer. Res. **9**, 269–316 (2009)
11. Van der Aalst, W.M.P.: A class of Petri net for modeling and analyzing business processes. Comput. Sci. Rep. **95**, 26 (1995)
12. Muehlen, M., Ho, D.T.-Y.: Risk management in the BPM lifecycle. In: Bussler, C.J., Haller, A. (eds.) BPM 2005. LNCS, vol. 3812, pp. 454–466. Springer, Heidelberg (2006). https://doi.org/10.1007/11678564_42
13. Caetano, A., Silva, A.R., Tribolet, J.: Using roles and business objects to model and understand business processes. In: Proceedings of the 2005 ACM Symposium on Applied Computing, pp. 1308–1313. ACM (2005)
14. Tyndale-Biscoe, S., Sims, O., Wood, B., Sluman, C.: Business modelling for component systems with UML. In: Proceedings of Sixth International Enterprise Distributed Object Computing Conference, EDOC 2002, pp. 120–131. IEEE (2002)
15. Born, M.: User Guidance in Business Process Modelling. Logos Verlag Berlin GmbH, Berlin (2012)
16. Vom Brocke, J., Mendling, J.: Business Process Management Cases. Digital Innovation and Business Transformation in Practice. Springer, Heidelberg, Berlin (2018). https://doi.org/10.1007/978-3-319-58307-5
17. Ma, Z., Leymann, F.: A lifecycle model for using process fragment in business process modeling. In: Proceedings of the 9th Workshop on Business Process Modeling, Development, and Support (BPDMS 2008), pp. 1–9 (2008)
18. van der Aalst, W.M.: Business process management: a personal view. Bus. Process Manag. J. **10** (2004). https://www.emeraldinsight.com/doi/abs/10.1108/bpmj.2004.15710baa.001?journalCode=bpmj
19. Ruževičius, J., Milinavičiūtė, I., Klimas, D.: Peculiarities of the business process management lifecycle at different maturity levels: the banking sector's case. Issues Bus. Law **4**, 69–85 (2012)
20. Macedo de Morais, R., Kazan, S., Inês Dallavalle de Pádua, S., Lucirton Costa, A.: An analysis of BPM lifecycles: from a literature review to a framework proposal. Bus. Process Manag. J. **20**, 412–432 (2014)
21. Bernardo, R., Galina, S.V.R., de Pádua, S.I.D.: The BPM lifecycle: How to incorporate a view external to the organization through dynamic capability. Bus. Process Manag. J. **23**, 155–175 (2017)
22. Netjes, M., Reijers, H.A., van der Aalst, W.M.: Supporting the BPM life-cycle with FileNet. In: Proceedings of the CAiSE, pp. 497–508. Citeseer (2006)

23. Reichert, M., Weber, B.: Enabling Flexibility in Process-Aware Information Systems: Challenges, Methods. Technologies. Springer, Heidelberg (2012). https://doi.org/10.1007/978-3-642-30409-5

24. Jesus, L., Rosemann, M.: The future BPM: seven opportunities to become the butcher and not the Turkey. BPTrends, February 2017

25. Leth, S.A.: Critical success factors for reengineering business processes. Natl. Prod. Rev. 13, 557–568 (1994)

26. vom Brocke, J., Zelt, S., Schmiedel, T.: On the role of context in business process management. Int. J. Inf. Manag. 36, 486–495 (2016)

27. Zhao, X., Mafuz, S.: Towards incorporating context awareness into business process management. World Acad. Sci. Eng. Technol. Int. J. Soc. Behav. Educ. Econ. Bus. Ind. Eng. 9, 3890–3897 (2015)

28. Le Vie, D.S., Donald, S.: Understanding data flow diagrams. In: Annual Conference - Society for Technical Communication, pp. 396–401 (2000)

29. Edwards, C.A.: Business process integrity and enterprise resource planning systems: an analysis of workaround practices in a large public sector organisation (2013). http://scholar.sun.ac.za/handle/10019.1/79845

30. Draheim, D.: Semantics of Business Process Models. In: Draheim, D. (ed.) Business process technology, pp. 75–117. Springer, Heidelberg (2010). https://doi.org/10.1007/978-3-642-01588-5_4

31. Bucher, T., Winter, R.: Taxonomy of business process management approaches. In: vom Brocke, J., Rosemann, M. (eds.) Handbook on Business Process Management 2, pp. 93–114. Springer, Heidelberg (2010). https://doi.org/10.1007/978-3-642-01982-1_5

32. Hallerbach, A., Bauer, T., Reichert, M.: Managing process variants in the process lifecycle (2008)

33. Governatori, G., Rotolo, A.: A conceptually rich model of business process compliance. In: Proceedings of the Seventh Asia-Pacific Conference on Conceptual Modelling, vol. 110, pp. 3–12. Australian Computer Society, Inc. (2010)

34. Ghanavati, S., Amyot, D., Peyton, L.: A systematic review of goal-oriented requirements management frameworks for business process compliance. In: 2011 Fourth International Workshop on Requirements Engineering and Law (RELAW), pp. 25–34. IEEE (2011)

35. Cabanillas, C., Resinas, M., Ruiz-Cortés, A.: Exploring features of a full-coverage integrated solution for business process compliance. In: Salinesi, C., Pastor, O. (eds.) CAiSE 2011. LNBIP, vol. 83, pp. 218–227. Springer, Heidelberg (2011). https://doi.org/10.1007/978-3-642-22056-2_24

36. Cabanillas Macías, C., Resinas Arias de Reyna, M., Ruiz Cortés, A.: Hints on how to face business process compliance. In: III Taller De Procesos De Negocio E Ingeniería De Servicios, PNIS2010, Valencia, España (2010)

37. Sadiq, S., Governatori, G., Namiri, K.: Modeling control objectives for business process compliance. In: Alonso, G., Dadam, P., Rosemann, M. (eds.) BPM 2007. LNCS, vol. 4714, pp. 149–164. Springer, Heidelberg (2007). https://doi.org/10.1007/978-3-540-75183-0_12

38. Daniel, F., et al.: Business compliance governance in service-oriented architectures. In: 2009 International Conference on Advanced Information Networking and Applications (AINA 2009), pp. 113–120. IEEE (2009)

How to Support Group Decision Making in Horticulture: An Approach Based on the Combination of a Centralized Mathematical Model and a Group Decision Support System

Pascale Zaraté[1](✉) ⓘ, MME Alemany[2] ⓘ, Mariana del Pino[3],
Ana Esteso Alvarez[2] ⓘ, and Guy Camilleri[1] ⓘ

[1] IRIT, Toulouse University, Toulouse, France
{zarate, camiller}@irit.fr
[2] CIGIP, Universitat Politècnica de València, Camino de Vera S/N,
46002 Valencia, Spain
mareva@omp.upv.es, aesteso@cigip.upv.es
[3] FCAyF UNLP, La Anunciación - Huerta Orgánica, La Plata, Argentina
mdelpino@agro.unlp.edu.ar

Abstract. Decision making for farms is a complex task. Farmers have to fix the price of their production but several parameters have to be taken into account: harvesting, seeds, ground, season etc.… This task is even more difficult when a group of farmers must make the decision. Generally, optimization models support the farmers to find no dominated solutions, but the problem remains difficult if they have to agree on one solution. In order to support the farmers for this complex decision we combine two approaches. We firstly generate a set of no dominated solutions thanks to a centralized optimization model. Based on this set of solution we then used a Group Decision Support System called GRUS for choosing the best solution for the group of farmers. The combined approach allows us to determine the best solution for the group in a consensual way. This combination of approaches is very innovative for the Agriculture domain.

Keywords: Centralized optimization model · Group decision support system · AgriBusiness

1 Introduction

Fixing the price of farms products is always a hard decision. The real food prices are determined by the food supply-demand balance [1]. The price to be determined is generally function on demand but also on supply [2]. Farmers usually select which crops to plant in function of the expected benefits that will be produced. Nevertheless, if all farmers decide to plant the same crops, this would result in a decrease of the crop's sales price, turning it less profitable. Simultaneously, the supply of less profitable crops would be lower than their demand, resulting in an increase of their final sales price and, therefore, in their conversion into more profitable crops. It is then

P. S. A. Freitas et al. (Eds.): EmC-ICDSST 2019, LNBIP 348, pp. 83–94, 2019.
https://doi.org/10.1007/978-3-030-18819-1_7

mandatory to effectively match demand and supply in the agri-food supply chain processes [3]. The remaining question is then, how can farmers decide which crops to cultivate each season to maximize their profits?

It has been proved by [4] that one solution to this problem could be to centrally plan the planting and harvest for all the farmers while maximizing the profits of the region. However, this solution could produce inequalities in the profits obtained by farmers, leading to the unwillingness to cooperate.

In this paper, we aim to prove that making decisions for farmers using profitable information can lead to a better global decision. To achieve this objective, we used two technics: one coming from mathematical modelling and one coming from the Group Decision Support Systems. It has been proved by [4] it is more favorable to reach an optimal solution for the whole supply chain and then, share it between its members; that implies that the profits obtained by farmers can be maximized and the inequalities between them can be reduced when centrally planning the planting and harvest of crops. A centralized optimal solution is then used in this paper as the best solution for this problem. It will be the benchmark of our study. This information is used in the group decision-making process.

We aim to show how a group engaged in a decision-making problem is influenced by the information that is available. For this purpose, we developed an experimental study. This study is based on the combination of two methodologies. We firstly generated a list of alternatives thanks to mathematical centralized model and then we used a Group Decision Support System. Our main goal is to combine two approaches to generate a satisfactory solution for a group. The paper is organized as follows. In the next section, we describe the related works on the two used technics, i.e. the GDSS and mathematical modeling for used for agriculture or horticulture purpose. The third section we present the used centralized mathematical model. In the fourth section briefly describes the used GDSS called GRoUp Support (GRUS) [5]. In the fifth section we describe the experiment decomposed by three subsections: 1. description of the used scenario, 2. presentation of the obtained alternatives by the centralized mathematical model and 3. description of the second GRUS use. In the sixth section, we analyze the obtained results and we conclude the paper in the last section.

2 Related Work

2.1 Group Decision Support Systems for Agriculture or Horticulture

GDSS are designed to support a group engaged in a decision-making process. There are a lot of study on group creativity and [6] reported a study that is descriptive in nature and designed to generate hypotheses that will form the basis for future research in order to facilitate group creativity. The used application domain is generally business oriented.

Some studies report the design of DSS for agriculture. Recent approaches in building decision support systems (DSS) for agriculture, and more generally for environmental problems, tend to adopt a "systemic" approach [7] focus on design issues faced during the development of a DSS to be used by technicians of the advisory service performing pest management according to an integrated production approach. These last studies report on systems designed for single user and not for a group of decision makers.

Nevertheless, decisions to make are also a question of group of persons in the Agriculture domain. For example, when the products are ready to be sent the supply chain process involves a group of stakeholders: farmers, sellers, transporters, auctions persons. There is a need to develop a process and a support for a group engaged in a decision-making process in agriculture.

2.2 Collaborative Planning for Agriculture or Horticulture

An increasing number of recent research works recognize the necessity of implementing collaboration mechanisms among the members of fruit and vegetable SCs for achieving sustainability [8], increase revenues and customer satisfaction and reduce the negative impact of uncertainty [9]. [10] distinguish three interrelated dimensions of collaboration: information sharing, decision synchronization, and incentive alignment. In the context of decision synchronization, we center on collaborative operations planning at the tactical level. Different literature reviews [11, 12] conclude the shortage of research addressing collaborative planning issues in the agricultural sector and the scarce number of integrated planning models. When collaborative planning is implemented under a distributed approach, it is necessary to implement coordination mechanisms [13]. [14] affirm that still, research on coordination-related issues in an agricultural supply chain is in its early development and not cover coordination of the whole supply chain. They state that studies on the coordination of processed fruits and vegetables products have been more widely studied than the coordination of fresh produce.

In their review, [14] also identify mathematical modelling as one methodology used in agri-food supply chain coordination. One application can be found in the work of [15] who propose a distributed mathematical model for the coordination of perishable crop production among small farmers and a consolidation facility using auction mechanisms. Another example is the research of [9] where a collaborative mathematical model is proposed to improve farmers' skill level by investments in an uncertain context.

[14] conclude in their review that studies on supply chain coordination in agri-food sector with a particular focus on small-scale farmers is very scarce. Besides, [16] highlight as a conclusion of their review that although quantitative modeling approaches have been applied to agricultural problems for a long time, adoption of these methods for improving planning decisions in agribusiness supply chains under uncertainty is still limited. [17] identify as new opportunities for operations research in agri-food SC better predictive modelling of the decision making behavior of actors in the natural resources system, multiple stakeholder decision analysis, optimization in a more complex business environment and multi-criteria decision making. [18] affirm that when dealing with the complexity of agri-food supply chain, sustainability is one of perspectives that can be applied to maintain the competitive strategies in economic, environmental, and social aspects that is called triple bottom line. For that, multi-criteria or multi-objective decision support tools should be developed that take into

account the three dimensions of sustainability. [19] propose hybrid-modelling approaches to cope with the complexity of real-world Sustainable Food SC in order to obtain managerial insights.

It can be drawn as a conclusion that research on coordination issues in agricultural SCs is in its early development. Moreover, research addressing coordination among actors in the same stage specifically at the farmer stage is even more scarce. In view of this, this paper analyses how the multi-criteria group decision-makingbehavior of small farmers supported by GRUS DSS is affected by the optimal solution knowledge obtained from a mathematical model. Three objectives (criteria) related to the economic, social and environmental categories are considered to achieve the sustainability of the horticulture supply chain coping, therefore, with the so-called triple bottom line. Therefore, with this work we contribute to fill the scarcity of works dealing with multiple stakeholder decision analysis, coordination among small farmers, predictive modelling of their decision-makingbehavior and application of hybrid modelling approaches to achieve the sustainability in horticulture SCs.

3   Mathematical Model for the Tomato Planning Problem

A mixed integer linear programming model has been developed to support the centralized decision making about: the time and quantity of different types of tomato to be planted and harvested by different farmers, the quantity of each type of tomato to be transported from the farmer to each market as well as the unfulfilled demand for each type of tomato and market. The main reason for defining two different decision variables for planting and harvesting quantities stems from the fact that planting and harvesting time periods are different. Therefore, it is important to detail not only how much is harvested but also when it is harvested and put on the market in order to match the market demand at prices as high as possible. Due to the yield of fields in each period is an uncontrollable variable by farmers, it could happen that the quantity ready to be harvested per period was higher than the market demand. In this scenario, the farmer could decide not to harvest all the tomatoes that have matured in order to save additional costs. Based on this, the quantity of each type of tomato wasted at each period in each farm is derived.

The optimum value for the above decision variables in the the supply chain will depend on the specific input data and the objectives pursued. As regards the input data, the following information is required: the estimation of the selling price and the market demand for the different types of tomato and for each time period, the yield for each farmer and tomato type, the density of cultivation, the total area available for planting in each farm, the activities to be carried for each type of tomato and the resources consumed, the costs of labor, waste, transporting tomatoes and unfulfilled demand. Feasible dates to plant and harvest each tomato type are also necessary.

When making the above decisions the three dimensions of SC sustainability are taking into account by the definition of three conflicting objectives that give rise to a multi-objective model. These objectives are the following:

- Economic Objective: The first objective consists in maximizing the profits of the whole supply chain calculated as the sales incomes minus the total costs. These costs contemplate those incurred due to tomatoes production in each farm and the distribution from each farm to each market.
- Environmental Objective: The second objective aims at minimizing the total waste along the Supply Chain. The maximum profit does not necessarily imply the minimum waste: a famer can decide to plant a quantity of tomatoes in some specific periods that allow him to sell some quantity of tomatoes in the season with the highest prices. But this decision, that can imply the maximum profit, can also imply more waste because of the uncontrollable yield distribution. Therefore, the profit maximization and the waste minimization can be conflicting objectives. Because the minimization of the food loss and waste is one of the environmental sustainable objectives recognized in several studies and organisms such as FAO [20], we have introduced this objective in our model.
- Social Objective: The third objective tries to minimize the unfulfilled demand along all the Supply Chain covering human requirements and increasing the customer satisfaction.

The decisions made should respect the following constraints. The acreage for each type of tomato should not exceed the available planting area in each farm. It is necessary to ensure that all tomato types are planted in all planting periods. At the same time, it is required that all farmers plant tomatoes at all planting periods to ensure the flow of products. The maximum quantity to be harvested at each period should not be higher than the yield per unit area harvested. It is not possible to transport from each farmer to each market tomato quantities higher than those harvested in the same farm for each time period. The waste in each farm is calculated as the difference between matured tomatoes and those not harvested or transported. The balance equation for calculating the unfulfilled demand for each type of tomato and market is based on the difference between the market demand for each tomato type and the total quantity of this type of tomato transported from all farmers to the market. If more product was transported to markets than the necessary one to fulfil the demand, the exceeding tomatoes were wasted. The quantity of tomatoes that was finally sold could not exceed the supply nor the demand. Constraints arc defined to ensure the coherence between the integer and binary variables related to the planting decision.

4 GRoUp Support (GRUS) Description

The GRUS (GRoUp Support) system is a Group Decision Support System (GDSS) in the form of a web application developed on the GRAILS framework (an open source platform). GRUS can be used for making collaborative meetings where all participants are connected to the system at the same time or at different time; in the same location (room) or in different locations. GRUS requires an internet connection and provides classical functionalities of multi-user web applications (sign in/sign out, user management, etc.). With GRUS, a user can participate to several meetings at the same time. She/he can facilitate (animate) some of them and only participate as a standard user to other ones.

The GRUS system is based on collaborative tools, the main tools are: electronic brainstorming tools, clustering tools, vote tools, multi-criteria tool, etc. A collaborative process in GRUS corresponds to a sequence of collaborative tools. A collaborative meeting requires one facilitator, which can always contribute to the meeting.

A GRUS meeting is composed of two general steps: the meeting creation and the meeting achievement. In the meeting creation step, a user (usually the facilitator) defines the topic of the meeting, the facilitator, the group process, the beginning date and the duration. The facilitator can reuse an existing group process or can define a new one (see Fig. 1). In the second step (meeting achievement), the facilitator manages the meeting thanks to a toolbar (see Fig. 2). This toolbar is only available in the facilitator interface; other participants do not have it and just follow the group process. With this toolbar, the facilitator can: add/remove participants, go to the next collaborative tool, modify the group process and finish the meeting.

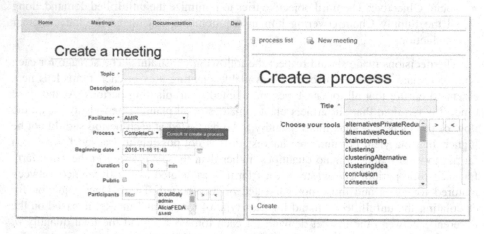

Fig. 1. Meeting and process creation

Fig. 2. On the left standard participant interface, on the right facilitator interface with the toolbar

5 Experiment

5.1 Scenario/Context

For the decision-making situation under study, we consider five farmers in the region of La Plata, Buenos Aires, Argentina, with an available planting area in hectare (ha) for each farmer of 20, 18, 17, 16 and 15, respectively. Our horizon is one year divided into monthly periods. Three types of tomatoes can be planted during three different months (July, October, and January) that do not depend on the specific type. The harvesting periods are the same for each type but depends on the planting period (Table 1). These planting periods are the usual in the region of La Plata, that is one of the most important areas of tomato in greenhouse for sell in fresh in Argentina.

Table 1. Harvesting periods

	07	08	09	10	11	12	01	02	03	04	05	06
July					X	X	X	X				
October							X	X	X	X		
January									X	X	X	X

During the growth of the plant from the planted date to the harvesting date, different activities need to be made to the plant in order to ensure its correct growth. These activities are called cultural practices. Each variety requires a different number of cultural practices at different time to perform each activity. Besides, one plant of each type of tomato can be harvested different number of times during the harvesting period and requires different time to harvest per plant. Both, the cultural practices and harvest activities, are made by laborers with limited capacity and with contracting costs.

The yield of the plant per month is dependent on the planting date and the type of tomato planted. The yield represents the kilograms (kg) of tomatoes that can be harvested per month from a single plant.

Once harvested the tomatoes are distributed to two different customers: a central market and some restaurants. The cost to transport one kg of tomatoes depends on the origin (farmers) and the destination (type of customer). The demand for each type of tomato is defined based on the month and market.

The price for each type of tomato also depends on the month in which it is sold. In addition, it is considered that sale prices vary in function of the balance between supply and demand. We estimate that prices increase when the total supply from all farmers is lower than demand. Prices decreases when the supply is higher than demand. In cases where some of the demand is not fulfilled because there is not enough supply (demand > supply), the benefit to be obtained is penalized with a cost. The penalization cost is calculated as ½ of the most probable price. Another penalization cost is included for cases in which some product is wasted throughout the supply chain (demand < supply). In its current state, the experiment does not take into account the fact that side payments would be possible to make the generated solution acceptable for all group members.

5.2   Results of the Centralized Mathematical Model

To solve the multi-objective model, we transformed it into a single-objective model by applying the ε-constraint method [21, 22]. In this method, one of the objectives is selected as the model's objective function, while the other objectives are considered the model's constraints. The right-hand side (RHS) of these constraints are defined by the grid points (εi) that are obtained by dividing the objective's ranges of values into as many equal intervals as desired. The ranges of values that each objective modelled as a constraint can assume are determined by a lexicographic optimization proposed by [22].

Following this method, the model is optimized for one objective. Then, the model is optimized for a second objective by constraining the value of the first objective to its optimal value. The same process is made with the third objective by constraining both the first and second objective. When repeating the process for the different combinations of the objectives, a set of solutions is provided. Dominated solutions are discarded and non-dominated solutions are analyzed to identify the best and worst values for each objective. These values define the range of values used to define the grid points. Once the model is run for the different grid points combinations, solutions obtained do not necessarily have to be equally distributed in the objective's values.

For our case study, ten values were defined for the εi parameter. The model was implemented using the MPL software 5.0.6.114 and the solver Gurobi 8.0.1. This provide us with ten non-dominated solutions. The detail for each non-dominated solution can be consulted in Table 2 of Annex I. For each solution, the value of the three objective functions for the entire supply chain and for each farmer are presented. The area of land dedicated to each type of tomato in each farm are also reported. As it can be checked for the solutions reported, the profit, wastes and unfulfilled demand for each farmer varies with solutions and a solution that reports the best objective function for one farmer can be the worst for the other ones. Consequently, it is necessary a complementary procedure to decide which non-dominated solution to implement. This procedure is described in the following section.

This model could also be used in a distributed way by reducing the number of farmers to one. Obtained non-dominated solutions would not be non-dominated for the whole supply chain but only for the particular farmer.

5.3   GRUS Experiment Using Solutions Generated by the Centralized Model

We used GRUS to rank the 10 generated alternatives. We were five decision makers playing the role of the farmers, including the facilitator as a decision maker. The adopted process was composed by three steps and was the following:

1. Alternatives Generation: The facilitator filled in the system the 10 solutions found thanks to the optimization model.
2. Vote: The five decision makers ranked the 10 solutions according to their own preferences.
3. The system then computes the final ranking for the group using the Borda [23] methodology.

The result is described in the Fig. 3.

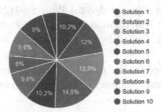

Fig. 3. Result of the group ranking. 1. Solution 4: 24 points, 2. Solution 3: 23 points, 3. Solution 2: 20 points, 4. Solutions 1 and 5: 17 points, 5. Solutions 6 and 8: 16 points, 6. Solution 9: 15 points, 7. Solution 7: 10 points and 8. Solution 10: 8 points

This result is given for the group of five farmers. The five farmers have the same weight (importance) for this experiment. Nevertheless, we also could choose that the importance of each farmer is linked to the number of hectares, only in Multi-Criteria processes.

We can see that on positions 4 and 5 two alternatives are ex aequo: solutions 1 and 5 for rank 4 and solutions 6 and 8 for rank 5. The best solution for the group is the one for which the five farmers have benefits and the three kinds of tomatoes are planted, that is solution number 4. Nevertheless, we can notice that it is not the solution, which generates the best profit on a global point of view.

This experiment shows that the solution obtained by a centralized optimization model that generates the highest profit, that is the solution 1 in the table of the Annex 1, is not necessarily the best one for the group of agents (humans).

6 Conclusion

In this paper, we combined two approaches in order to generate a good solution for a group of human beings. The application domain is the Agriculture. Planning a strategy of production is a difficult task in the agriculture if several constraints, like for example harvesting, ground to plant, choose the best seed, etc. are taken into account.

First of all, we generated 10 solutions thanks to a centralized optimization model. These solutions are then explained to the group of five farmers. We, in a second step, asked to the five farmers to give their own preferences on these 10 solutions. We finally used a Group Decision Support System, called GRUS, to find the final ranking for the group. This final ranking is based on the preferences given by the stakeholders. Nevertheless, the conclusions of this experiment have some limitations based on the fact the decision makers were researchers and not farmers. We still need to do the same experiment with real farmers and obtain their feedback about the process.

We show in this paper how the GDSS GRUS is helpful to generate a group decision which reduces conflicts in a group (Borda voting procedure) and how it supports to find a consensus. These results are interesting but we need to conduct more experiments with a decentralized optimization model and compare the obtained non-dominated solutions with the solutions obtained with the GRUS system.

Acknowledgment. The authors acknowledge the Project 691249, RUC-APS: Enhancing and implementing Knowledge based ICT solutions within high Risk and Uncertain Conditions for Agriculture Production Systems, funded by the EU under its funding scheme H2020-MSCA-RISE-2015.

One of the authors acknowledges the partial support of the Programme of Formation of University Professors of the Spanish Ministry of Education, Culture, and Sport (FPU15/03595).

ANNEX 1

Table 2. Set of non-dominated optimal solutions for the mathematical programming model

Solution	Profits (€) SC	#	Profits (€) Farm	Tomato wastes (kg) SC	Tomato wastes (kg) Farm	Unmet demand	Cherry tomato planting area (ha) SC	Cherry tomato planting area (ha) Farm	Round tomato planting area (ha) SC	Round tomato planting area (ha) Farm	Pear tomato planting area (ha) SC	Pear tomato planting area (ha) Farm
1	148.334.625	1	24.758.476	5.316.020	998.708	207.317.999	35,9365	17,9365	21,6970		28,3665	2,0635
		2	21.892.373					18,0000				
		3	39.408.112							11,6278		5,3722
		4	32.890.933		4.317.312					1,5670		14,4330
		5	29.384.732							8,5023		6,4977
2	148.302.280	1	25.086.408	5.315.998	2.115.428	201.749.612	33,6292	15,6292	24,0044		28,3665	4,3708
		2	21.891.207					18,0000				
		3	39.407.029							11,6277		5,3723
		4	34.825.732		3.200.570					3,3830		12,6170
		5	27.091.904							8,9937		6,0063
3	148.003.481	1	25.818.920	6.417.520	3.958.788	195.841.392	30,4959	12,4959	26,0250	1,1288	29,4791	6,3753
		2	21.889.971					18,0000				
		3	39.405.833		12					11,6277		5,3723
		4	35.569.237		2.458.720					4,2743		11,7257
		5	25.319.522							8,9941		6,0059
4	146.849.751	1	26.249.394	11.193.326	8.734.549	189.933.239	25,6717	7,6717	26,0250	1,1293	34,3032	11,1989
		2	21.888.734					18,0000				
		3	39.404.693							11,6277		5,3723
		4	35.568.111		2.458.765					4,2743		11,7257
		5	23.738.819		12					8,9937		6,0063
5	145.326.260	1	23.810.235	14.017.213	11.558.336	184.025.050	21,0899	3,0900	26,3350	1,4393	38,5751	15,4707
		2	21.887.500		22			17,9999				
		3	39.403.535		10					11,6277		5,3723
		4	35.566.937		2.458.822					4,2743		11,7257
		5	24.657.938		24					8,9936		6,0064
6	142.518.888	1	23.757.449	11.213.768	8.754.980	178.116.854	18,7100	0,7100	31,3691	6,4735	35,9209	12,8165
		2	21.886.261					18,0000				
		3	39.402.357							11,6277		5,3723
		4	35.565.913		2.458.765					4,2743		11,7257
		5	21.906.908		23					8,9936		6,0064
7	136.863.913	1	15.839.594	8.410.330	4.466.454	172.208.666	14,4576		33,0921	7,3206	38,4503	12,6794
		2	22.373.720		1.714.531			14,4576				3,5424
		3	39.401.183							11,6277		5,3723
		4	35.435.025		2.229.345					5,1500		10,8500
		5	23.814.391							8,9938		6,0062
8	146.572.577	1	25.244.207	-		204.769.167	37,8087	17,6497	26,0250		22,1661	2,3503
		2	21.891.837					18,0000				
		3	39.479.894							11,9768		5,0232
		4	34.626.196					2,1591		3,9252		9,9157
		5	25.330.443							10,1230		4,8770
9	135.083.010	1	22.220.586	-		182.724.221	22,8945	2,7249	32,5065	3,5312	30,5989	13,7439
		2	21.887.918					17,9982		0,0016		0,0001
		3	39.979.961							7,3755		9,6245
		4	34.100.105					2,1714		8,5266		5,3020
		5	16.894.441							13,0716		1,9284
10	129.129.328	1	15.544.979	25.230.996	8.427.387	154.484.078	0,0004	0,0003	39,2378	8,9728	46,7618	11,0269
		2	19.246.325		14.995.650			0,0001		2,8857		15,1142
		3	39.397.689							11,6277		5,3723
		4	35.193.389		1.807.927					6,7579		9,2421
		5	19.746.946		32					8,9937		6,0063

References

1. Tweeten, L., Thompson, S.: Long-term global agricultural output supply-demand balance. Farm Policy J. **6**(1), 1–16 (2009)
2. Weintraub, A., Romero, C.: Operations research models and the management of agricultural and forestry resources: a review and comparison. Interfaces (Provid.) **36**, 446–457 (2006). https://doi.org/10.1287/inte.1060.0
3. Alemany, M.M.E., Grillo, H., Ortiz, A., Fuertes-Miquel, V.S.: A fuzzy model for shortage planning under uncertainty due to lack of homogeneity in planned production lots. Appl. Math. Model. **39**, 4463–4481 (2015). https://doi.org/10.1016/j.apm.2014.12.057
4. Stadtler, H.: A framework for collaborative planning and state-of-the-art. OR Spectr. **31**, 5–30 (2009). https://doi.org/10.1007/s00291-007-0104-5
5. Zaraté, P., Kilgour, D.Marc, Hipel, K.: Private or common criteria in a multi-criteria group decision support system: an experiment. In: Yuizono, T., Ogata, H., Hoppe, U., Vassileva, J. (eds.) CRIWG 2016. LNCS, vol. 9848, pp. 1–12. Springer, Cham (2016). https://doi.org/10.1007/978-3-319-44799-5_1
6. Nunamaker Jr., J.F., Applegate, L.M., Konsynski, B.R.: Facilitating group creativity: experience with a group decision support system. J. Manag. Inf. Syst. **3**(4), 5–19 (1987)
7. Perini, A., Susi, A.: Developing a decision support system for integrated production in agriculture. Environ. Model Softw. **19**(9), 821–829 (2004)
8. Dania, W.A.P., Xing, K., Amer, Y.: Collaboration behavioural factors for sustainable agri-food supply chains: a systematic review. J. Clean. Prod. **186**, 851–864 (2018)
9. Esteso, A., Alemany, M.M.E., Ortiz, A.: Conceptual framework for designing agri-food supply chains under uncertainty by mathematical programming models. Int. J. Prod. Res. **56**(13), 4418–4446 (2018)
10. Simatupang, T.M., Sridharan, R.: The collaboration index: a measure for supply chain collaboration. Int. J. Phys. Distrib. Logist. Manag. **35**(1), 44–62 (2005)
11. Ahumada, O., Villalobos, J.R.: Application of planning models in the agri-food supply chain: a review. Eur. J. Oper. Res. **196**(1), 1–20 (2009)
12. Tsolakis, N.K., Keramydas, C.A., Toka, A.K., Aidonis, D.A., Iakovou, E.T.: Agrifood supply chain management: a comprehensive hierarchical decision-making framework and a critical taxonomy. Biosyst. Eng. **120**, 47–64 (2014)
13. Alemany, M.M.E., Alarcón, F., Lario, F.C., Boj, J.J.: An application to support the temporal and spatial distributed decision-making process in supply chain collaborative planning. Comput. Ind. **62**(5), 519–540 (2011)
14. Handayati, Y., Simatupang, T.M., Perdana, T.: Agri-food supply chain coordination: the state-of-the-art and recent developments. Logist. Res. **8**(1), 1–15 (2015)
15. Mason, A.N., Villalobos, J.R.: Coordination of perishable crop production using auction mechanisms. Agric. Syst. **138**, 18–30 (2015)
16. Behzadi, G., O'Sullivan, M.J., Olsen, T.L., Zhang, A.: Agribusiness supply chain risk management: a review of quantitative decision models. Omega (U. K.) **79**, 21–42 (2018)
17. Plà, L., Sandars, D., Higgins, A.: A perspective on operational research prospects for agriculture. J. Oper. Res. Soc. **65**, 1078–1089 (2014)
18. Dania, W.A.P., Xing, K., Amer, Y.: Collaboration and sustainable agri-food suply chain: a literature review. In: MATEC Web Conference, vol. 58 (2016)
19. Zhu, Z., Chu, F., Dolgui, A., Chu, C., Zhou, W., Piramuthu, S.: Recent advances and opportunities in sustainable food supply chain: a model-oriented review. Int. J. Prod. Res. **7543**, 1–23 (2018)

20. Porata, R., Lichtera, A., Terryb, L.A., Harkerc, R., Buzbyd, J.: Postharvest losses of fruit and vegetables during retail and in consumers' homes: quantifications, causes, and means of prevention. Postharvest Biol. Technol. **139**, 135–149 (2018)
21. Ehrgott, M.: Multicriteria Optimization. Springer, Heidelberg (2005)
22. Mavrotas, G.: Effective implementation of the ε-constraint method in multi-objective mathematical programming problems. Appl. Math. Comput. **213**(2), 455–465 (2009)
23. «Decision Maker» de Borda Institute. http://www.decision-maker.org/content/voting-systems. Accessed Jan 2017

Advances in Decision Support Systems' Methods and Technologies

Intelligent Multicriteria Decision Support System for a Periodic Prediction

Sarra Bouzayane[1]([⊠]) and Ines Saad[1,2]

[1] MIS Laboratory, University of Picardie Jules Verne, 80039 Amiens, France
{sarra.bouzayane,ines.saad}@u-picardie.fr
[2] Amiens Business School, 80039 Amiens, France

Abstract. This paper proposes an intelligent decision support system for the Incremental Periodic Prediction of the decision class to which an action is likely to belong. This method is based on three phases. The first consists of three steps: the construction of a family of criteria for the characterization of actions; the construction of a representative learning set for each of the decision classes; and the construction of a decision table. The second phase is based on the *DRSA-Incremental* algorithm that we propose for the inference and the updating of the set of decision rules following the sequential increment of the learning set. The third phase is meant to classify the "Potential Actions" in one of the predefined decision classes using the set of inferred decision rules. Our method is based on the DRSA (Dominance-based Rough Set Approach) which is a supervised learning technique permitting to extract the preferences of decision makers for the actions categorization. Applied in the context of MOOCs (Massive Open Online Courses) for the categorization of learners' profiles, our approach proved a satisfactory classification quality that reaches 0.66.

Keywords: Intelligent decision support system ·
Multicriteria decision making · Incremental prediction · MOOC

1 Introduction

The amount of data collected and stored into databases has significantly increased. Consequently, the traditional data analysis techniques have become unsuitable for the processing of such huge volumes of data and new techniques have emerged [9]. In this work, we focus on the machine learning techniques that were mainly used for the prediction or the classification issues.

Prediction is the reasoning from the current events to the future properties and events they will cause [20]. It is frequently based on the conventional techniques of machine learning such as the Markov chains [19] and the neural network [21]. However, these techniques are rather based on the size of the learning sample without considering the quality of the assignment examples it contains. This learning sample is usually chosen randomly, a thing that may affect the classification model quality.

© Springer Nature Switzerland AG 2019
P. S. A. Freitas et al. (Eds.): EmC-ICDSST 2019, LNBIP 348, pp. 97–110, 2019.
https://doi.org/10.1007/978-3-030-18819-1_8

In this work we propose a prediction method based on the DRSA (Dominance-based Rough Set Approach) [11] approach which is a multicriterion classification approach that relies on the rough set theory [17]. The DRSA approach helps decision makers infer decision rules from the assignments of examples of "Actions of reference" in decision classes based on the multiple decision makers' viewpoints. Our prediction method is called *MAI2P* (Multicriteria Approach for the Incremental Periodic Prediction) and it permits to characterize the actions according to their profiles and behaviors. It is based on our *DRSA-incremental* algorithm to deal with the prediction issue in the case of a permanently evolving learning set.

The approach *MAI2P* relies on the human decision makers preferences and consists of three phases: The construction of a decision table, the induction and the update of a set of decision rules called preference model, and the early classification (prediction) of the "Potential Actions" using the previously inferred decision rules. The first phase consists of three steps that are the construction of a training set called "actions of reference", the construction of a coherent criteria family, and the classification of each action in one of the predefined decision classes. The first and the second phases run at the end of the current period and the third phase runs at the beginning of the next period. The approach *MAI2P* has been applied and validated in the context of MOOCs (Massive Open Online Courses).

The paper is organized as follows: Sect. 2 discusses the related work. Section 3 presents the background. Section 4 introduces the approach *MAI2P* proposed for the periodic prediction. Section 5 presents the functional architecture of the decision support system. Section 6 presents a case study and discusses the corresponding results. Section 7 concludes the paper and advances some prospects.

2   Related Work

This section presents the machine learning techniques that are proposed to predict or to classify the learners profiles or the dropout rate in the context of MOOCs.

Balakrishnan proposed a model to predict the student' retention in MOOCs using two kinds of Hidden Markov Model (HMM) techniques; HMM with a single feature and HMM with multiple features. Prediction is based on the cumulative percentage of the available lecture videos watched, the number of threads viewed on the forum, the number of posts made on the forum and the number of times the course progress page was checked. This model has the goal of predicting the dropout of a student for the next week of the MOOC on the basis of his data for the current week. Experiments showed that the multiple features HMM gives more reasonable results than the single feature one. In addition, the percentage of the available lecture videos watched is the most efficient when using the single feature HMM [1]. Authors in [6] proposed a model based on the Neural Network to predict a student's attrition in MOOCs. And other than the classical attributes, such as the number of clicks made by the learner and the weekly number of the forum and the course pages viewed, authors integrated a sentiment score attribute. This is calculated using a lexicon-based sentiment analysis of the forum posts. Authors proved that the analysis of the sentiment expressed by students during their forum posts is an important indicator of the dropout behavior intention. This model permits to estimate whether the student will drop the course in the next week of the MOOC based on his data of the previous week. Authors in [7] proposed a model to predict the situation when the MOOC instructor has to intervene in the forum threads. The purpose is to help students get an answer, when necessary, from the instructor in order not to drop the course out. This work considers the posts of a thread as a chain of events and uses the Linear Chain Markov Model and the Global Chain Model. Each of these models takes as inputs both the features about the thread and those about each post in the thread. The thread structure and the lexical analysis of the posts are also considered and proved to have a positive impact on the prediction result. Authors in [14] proposed a model to predict what they called the "leader learners" in the MOOC. To do this, they used the Support Vector Machine as well as the language accommodation measure. Their method is based on the lexical analysis of the forums posts in order to identify the students by whom the language of the struggling students is influenced. The students whose language influences positively the other students are called "leaders of the struggling students" and will be mobilized to answer their questions on the forums to encourage them not to drop. Authors in [24] proposed a temporal modeling approach to predict students who are at risk to dropout the MOOC using the General Bayesian Network and the Decision Tree. The used features are the number of discussion posts, the number of forum views, the number of quiz views, the number of module views and the degree of the social network. Authors showed the importance of using the appended features input and applied the Principle Component Analysis to identify the breakpoint for turning off the features of the previous weeks. This model permits to predict in a chronological order over weeks the dropout behavior of students.

In this context of MOOCs, the models don't permit to properly characterize the learners profiles and randomly select the learning set. Thus, the method based-DRSA that we propose in the reminder of this paper overcomes this problem.

3   Background: Dominance-Based Rough Set Approach

DRSA was proposed by Greco et al. [11] and was inspired from the Rough Sets Theory [17]. It allows to compare actions through a dominance relation and takes into account the preferences of a decision maker to extract a preference model resulting in a set of decision rules. According to DRSA, a data table is a 4-tuple $S = \langle A, F, V, f \rangle$, where:

- A is a finite set of reference actions,
- F is a finite set of criteria,
- $V = \cup_{g \in F} V_g$ is the set of the possible values of criteria, and
- f is an information function $f : A \times F \longrightarrow V$ such that $f(x, g) \in V_g, \forall x \in A, \forall g \in F$.

F is often divided into a subset $C \neq \varnothing$ of condition attributes and a subset $D \neq \varnothing$ of decision attributes such that $C \cup D = F$ and $C \cap D = \varnothing$. In multicriteria decision making, the scale of criteria should be ordered according to a decreasing or an increasing preference of a decision maker. We also assume that the decision attribute set $D = \{d\}$ is a singleton that partitions A into a finite number of decision classes $Cl = \{Cl_n, n \in \{1, \ldots, N\}\}$. Furthermore, we suppose that the classes are preference-ordered, i.e., for all $r, s \in \{1, \ldots, N\}$ such that $r > s$, actions from Cl_r are preferred to actions from Cl_s.

Example: If we apply these notions for learners classifications in a context of MOOCs (Massive Open Online Courses) where actions are the learners and decision makers are the members of the pedagogical team, we can consider three decision classes: Cl_1 of the "At-risk learners", Cl_2 of the "Struggling learners" and Cl_3 of the "Leader learners". Thus F is divided into $D = \{1{:}Cl_1, 2{:} Cl_2, 3{:} Cl_3\}$ and a set C of condition criteria allowing the learners characterization as for example $g_1 =$ "Study level" and $g_2 =$ "Previous experience with MOOCs". The possible values of the criteria g_1 is $V_{g_1} = \{1{:}$ Scholar student; $2{:}$ High school student; $3{:}$ PhD Student; $4{:}$ Doctor$\}$.

Among the parameters that the approach DRSA defines, we cite:

Dominance Relation. Let $P \subseteq C$ be a subset of condition criteria. The dominance relation D_p associated with P is defined for each pair of actions x and y thus:

$$\forall(x,y) \in A, \ x \, D_p y \Leftrightarrow f(x, g_j) \succ f(y, g_j) \ \forall g_j \in P$$

To each action $x \in A$, two sets are associated:

- P-Dominating set $D_P^+(x) = \{y \in A : y D_p x\}$ containing actions that dominate x.
- P-dominated set $D_P^-(x) = \{y \in A : x D_p y\}$ containing actions dominated by x.

These sets are used to approximate decision classes. In the DRSA, the represented knowledge is a collection of downward unions $Cl_{\overline{n}}^{\leq}$ and upward unions $Cl_{\overline{n}}^{\geq}$ of decision classes. The assertion "$x \in Cl_{\overline{n}}^{\leq}$" means that "$x$ belongs to at most to the class Cl_n", while "$x \in Cl_{\overline{n}}^{\geq}$" means that "$x$ belongs to at least to the class Cl_n".

Lower Approximation. The P-lower approximations of $(Cl_{\overline{n}}^{\leq})$ and $(Cl_{\overline{n}}^{\geq})$ with respect to $P \subseteq C$ are respectively denoted $\underline{P}(Cl_{\overline{n}}^{\leq})$ and $\underline{P}(Cl_{\overline{n}}^{\geq})$. The P-lower approximation contains all actions that are assigned with certainty to an union of decision classes.

Decision Rule. A decision table may be considered as a set of "*if...then...*" decision rules, where the condition part specifies values assumed by one or more condition criteria and the decision part specifies an assignment to one or more decision classes. Decision rules used in this paper are represented as follows:

- **if** $f(x, g_1) \geq r_1 \wedge \ldots \wedge f(x, g_m) \geq r_m$ **then** $x \in Cl_{\bar{n}}^{\geq}$; such that $(r_1 \ldots r_m)$ $\in (v_{g_1} \ldots v_{g_m})$.
- **if** $f(x, g_1) \leq r_1 \wedge \ldots \wedge f(x, g_m) \leq r_m$ **then** $x \in Cl_{\bar{n}}^{\leq}$; such that $(r_1 \ldots r_m)$ $\in (v_{g_1} \ldots v_{g_m})$.

Example: we consider the decision rule presented as follows: $If f(A, g_1) \geq$ 4 *then* $A \in Cl_3$. This rule means that if a learner is at least a doctor he is a "leader learner" and will be classified at least in the decision class Cl_3.

4   Intelligent Decision Support System for the Incremental Prediction

MAI2P is a periodic prediction approach. It aims at the early classification of a set of actions in unions of predefined decision classes. This approach *MAI2P* consists of three phases: the construction of the decision table, the inference of a set of decision rules and the classification of each "Potential action" into one of the N predefined decision classes, using the previously inferred decision rules (cf. Fig. 1).

Phase 1: Construction of the Decision Table. This phase comprises three steps: The construction of a learning set of actions representative for each of the N decision classes, the construction of a coherent family of criteria to actions characterization and the construction of the decision table corresponding to the decision maker's preferences.

Step 1.1: The construction of a set of "Actions of reference". Data processing becomes more and more complicated when faced with a growing mass of data. Therefore, it is necessary to define a training sample containing a number

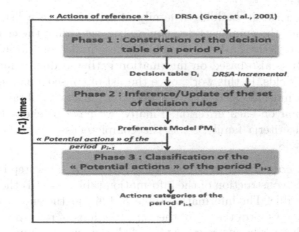

Fig. 1. Periodic multicriteria classification method (T is the prediction periods number)

of representative examples for each of the predefined decision classes Cl_n such that $n \in \{1..N\}$. In order to respect the terminology used in the DRSA approach, we call the training sample "Actions of reference". The training examples are selected by decision makers who must be experts in the decision domain.

Since the construction phase of the decision table strongly depends on the human dimension, the sample size must be taken into account. In effect, from a psychological viewpoint [13], a human decision maker is characterized by a cognitive capacity representing the upper limit to which he can associate his responses to the stimuli that are granted to him. Thus, when construction the training set, we must respect the quality of the selected actions, regardless of their size. In other words, the DRSA approach is based on the expertise of human decision makers who are experts in their fields in order to to build a sample of high quality and reasonable size. Thus, to classify new cases, DRSA uses new estimates by experts instead of using some previous cases with known results. This allows decision makers understanding and explaining the classifications made and so reviewing their previous decisions in case of hesitation.

The approach *MAI2P* must deal with the sets of non-stable "Actions of reference" that periodically vary throughout the prediction process. Thus, during each period P_i, the decision maker must select a new set A'_i of "Actions of reference" which is appended to all of the "Actions of reference", A_{i-1}, of all the previous periods. The set of "Actions of reference" of the period P_i is thus $A_i = A'_i + A_{i-1}; \forall i \in \{2..T\}$ such that T is the number of periods constituting the prediction process.

Step 1.2: The construction of a coherent family of criteria. A family of evaluation criteria must provide the judgment tools for a decision maker. It must check the coherence between his choices and expectations and the needs of the concerned actors [3]. Compared to an attribute, a criterion must be used to measure the preferences of a decision maker from a personal viewpoint [16]. The decision makers have a major role in the construction process of a coherent family of criteria.

This step consists in building a family of criteria from a list of indicators that can influence the opinion of decision makers regarding the characterization of actions. In multicriteria decision making these indicators are called "Consequences" [16]. It is also based on information gathered during interviews with experts in their fields. In this work, once the list of consequences is established, direct meetings must be held with the decision makers to obtain their preferential information on each criterion. Finally, we have to check the coherence of the obtained criteria family, so as the completeness, the cohesion and the non-redundancy [8].

Step 1.3: The construction of the decision table. This step is made of two sub-steps: (i) the construction of the information table, and (ii) the construction of the decision table. The information table S_i built at the end of a period P_i is a matrix whose rows form the set of the "m" "Actions of reference" identified in step 1.1 and whose columns represent the "p" criteria constructed in step 1.2. This matrix contains the evaluation function $f_i(A_{j,i}, g_k)$ of each action $A_{j,i} \in A'_i$

on each criterion $g_k \in F_1$ such that $i \in \{1..T\}$, $j \in \{1..m\}$ and $k \in \{1..p\}$. Variables T, m and p are respectively the number of periods, the size $|A_i'|$ of the "Actions of reference" set defined in the ith period and the size $|F_1|$ of the criteria family.

Analogously, variables $A_{j,i}$ and g_k constitute respectively the jth "Action of reference" in the set A_i' and the kth criterion. We remind that A_i' and F_1 represent respectively the set of "Actions of reference" constructed in the ith period and the family of criteria identified at the beginning of the decision process. Once the information table S_i is complete at the end of the ith period, we have to build the decision table with the expert decision makers during some meetings. Thus, we only have to add a column to the end of the information table. The last column concerns the affectation of each action of reference in one of the predefined decision classes (cf. Table 1). The decisions made by the decision makers about the classification of each "Action of reference" should be based on his/her assessment values on the set of all criteria. We call $D_j = \{d_{1,j}, d_{2,j}, .., d_{m,j}\}$ the decision vector.

Table 1. Example of a decision table built at the $Period_j$

	g_1	$\cdots$	g_k	$\cdots$	g_p	$\mathbf{D}_j$
$A_{1,j}$	$f(A_{1,j}, g_1)$	$\cdots$	$f(A_{1,j}, g_k)$	$\cdots$	$f(A_{1,j}, g_p)$	$d_{1,j}$
$A_{2,j}$	$f(A_{2,j}, g_1)$	$\cdots$	$f(A_{2,j}, g_k)$	$\cdots$	$f(A_{2,j}, g_p)$	$d_{2,j}$
$\cdots$	$\cdots$	$\cdots$	$\cdots$	$\cdots$	$\cdots$	
$A_{m,j}$	$f(A_{m,j}, g_1)$	$\cdots$	$f(A_{m,j}, g_k)$	$\cdots$	$f(A_{m,j}, g_p)$	$d_{m,j}$

Phase 2: Inference of a Preference Model Based on the Incremental Update of the DRSA Approximations. The preference model is a set of decision rules that permits to classify each action in one of the defined decision classes. It is inferred by applying an induction algorithm that takes as input the DRSA approximations (cf. Sect. 3). In this work, we used the rules induction algorithm *DOMLEM* proposed by the approach DRSA [11]. This algorithm generates a minimal set of decision rules covering all the examples in the decision table. This phase runs at the end of each period P_i during the prediction process and takes as input the decision table build in "Phase 1". This phase is made of two steps: The first consists in updating the DRSA approximations by applying our algorithm *DRSA-Incremental* [5]. The second step infers a preference model by applying the algorithm *DOMLEM* [11] on the updated approximations.

As quoted above, this approach deals with the case of "non- stable" set of "Reference actions". Thus, the decision support system should infer a preference model before all information is available and update it once a new portion of information arrive. According to [12], this situation of sequential flow of information is a primordial reason to incrementally update the DRSA approximations. Hence, instead of recalculating the approximations from scratch, we have to update them using our *DRSA-Incremental* algorithm to minimize the computational time.

The updated approximations will be provided as an input to the algorithm *DOMLEM*. Then, this algorithm provides a set of decision rules. The preference model of the period P_i represents the input of the phase 3 that will be used to classify the "Potential actions" at the beginning of the period P_{i+1}.

Phase 3: Classification of the "Potential actions". The third phase uses the previously inferred decision rules to assign each of the "Potential actions" to one of the N predefined decision classes.

A "potential action" is defined as a generic term used to describe an action or a referent of a decision. An action is considered potential, if it can be implemented or simply if it is considered "fertile" in a decision support context. The notion of potential action clarifies the nature of what constitutes the decision problem and formalizes the decision purpose. Thus, the set of potential actions represents the actions that can be classified in one of the predefined decision classes. In contrast to the learning set, the set of potential actions is usually large and encompasses all the informative actions that need categorization without resorting to prior selection.

This approach runs periodically: the first and the second phases run at the end of each period P_i such that i $\in \{1..T-1\}$ while the third phase runs at the beginning of each period P_i such that i $\in \{2..T\}$. The three phases are chained in a way that each phase inputs the output of the previous one (cf. Fig. 1).

5   Functional Architecture of the Decision Support System

The functional architecture we propose for this intelligent decision support system defines its operating principle.

The decision process is started by the definition of decision classes by both the Human study and the decision maker who must be expert in his domain. The human study is an actor who plays the role of mediator to help the decision makers make in their decision process, but must not in any way influence their decisions [18]. This step mainly relies on the analysis of data, that are usually stored in log or CSV files, using the different processing algorithms in the model base. These files are also used by the human study and the decision maker in order to collectively build the family of criteria to characterize the profile and behaviour of the actions. They have to agree on the appropriate aggregation algorithm to aggregate the criteria to be retained and to monitor their redundancy. Finally, the algorithm *DOMLEM* and *DRSA-Incremental* are stored in the model base in order to respectively infer and update the decision rules (Fig. 2).

The induction algorithm *DOMLEM* inputs the DRSA approximations to infer a preference model MP_i resulting in the set of decision rules. The preference

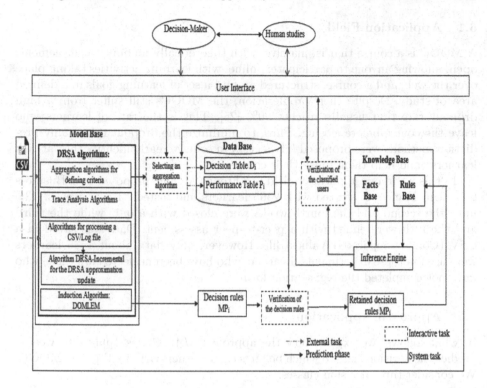

Fig. 2. Architecture of the intelligent decision support system

model must be verified by the human study and the decision maker in order to validate it or to select a consistent subset.

In the data base are stored the decision and the performance tables that gather the criteria, the actions evaluations on each criterion and the assignment decision. These data will be used for the actions classification phase that is based on DRSA approach.

The rules inferred by applying the *DOMLEM* and the *Incremental-DRSA* algorithms are stored in the knowledge base and applied on the performance table that is already transformed to facts. Thus, the inference engine should provide the final assignment of all potential actions. The content of the decision classes has to be checked by the decision maker in order to validate or to modify it.

6   Case Study

This section illustrates the approach *MAI2P* through a case study in the context of MOOC. We first introduce the application field. Then, we provide a step-by-step application of the approach *MAI2P*. Finally, we discuss the obtained results.

6.1 Application Field

A MOOC is a course that is, massive, with theoretically no limit to enrolement; open, allowing anyone to participate; online, with learning activities taking place over the web; and a course, structured around a set of learning goals in a defined area of study. Despite their proliferation, the MOOCs still suffer from a high dropout rate that usually reaches 90% [25]. This is the rate of learners who leave the course before closing. Thus, to minimize the dropout rate, many prediction methods were proposed (cf. Sect. 2) in order to early identify the "at-risk learners" and help them finish their course.

In this work, our application field is a French MOOC proposed by a Business School in France. It started with 2565 learners and lasted "T = 5" weeks. The first, the second and the fourth weeks were closed with a quiz while the third and the fifth were ended with a peer-to-peer assessment. Data was saved in a CSV (Comma-Separated Values) file. However, only data about 1535 learners were used in these experiments. Learners who have been neglected are those who have not completed the registration form.

6.2 Approach Application

In what follows, we explain how the approach $MAI2P$ was applied to weekly predict the decision class to which belongs each learner participating in a MOOC. We consider three decision classes:

- Cl_1. The decision class of the "At-risk learners" corresponding to learners who are likely to dropout the course in the next week of the MOOC.
- Cl_2. The decision class of the "Struggling learners" corresponding to learners who have some difficulties but are still active on the MOOC environment and don't have the intention to leave it at least in the next week of the MOOC.
- Cl_3. The decision class of the "Leader learners" corresponding to learners who are able to lead a team of learners by providing them with an accurate response.

Let $W = \{W_1, \dots W_i, \dots W_T\}$ be the set of weeks making up a MOOC such that $T \geq 2$ is the number of weeks a MOOC holds and W_i is the ith week of the MOOC. The three phases was applied as follows:

Phase 1: Construction of the Decision Table.

- *Step 1.1: The construction of a set of "Learners of reference".* Since the learners within a MOOC can enter or drop it at any time while it is running, the learning sample can not be stable over many weeks. Thus, at the end of each week W_i of the MOOC, the pedagogical team defined a new set A'_i of 30 "Learners of reference" that was directly appended as an additional set to the sets of "Learners of reference", A_{i-1}, of the previous weeks of the MOOC. And so, the set of "Learners of reference" of the week W_i becomes $A_i = A'_i + A_{i-1}, \forall i \in \{1..4\}$.

- *Step 1.2: The construction of a coherent family of criteria.* First, we identified a list of indicators that would permit to characterize a MOOC learner. These indicators may be static or dynamic and would give sign about the learner's skills, profile and motivation. Among the indicators mentioned in literature, we quote the study level [15], the MOOC language mastery and the motivation to participate in the MOOC [2], the cultural background [22], the level of technical skills and the lack of time [10]. Experts identify also the dynamic data which are traced according to the learner's activity. Authors in Wolff et al. [23] distinguished three types of activities permitting to caracterize a learner that are the access to a resource or to a course material; the publishing of a message on the forum and the access to the evaluation space. Second, to validate a final family of criteria, some meetings have been held with the pedagogical team. The pedagogical team constructed 11 criteria and applied a preference order on each of them. For example, the "study level" is an attribute that was identified in the literature [15] as an indicator allowing the learner evaluation in a MOOC context. Then, following meetings with the pedagogical team we were able to elicit their preferences on this attribute (1: Scholar student; 2: High school student; 3: PhD Student; 4: Doctor) to obtain a criterion. This step and the final criteria family are detailed in [4].
- *Step 1.3: The construction of the decision table.* At the end of each week W_i of the MOOC, a matrix, whose rows form the set of the 30 "Learners of reference" and whose columns represent the 11 evaluation criteria, is built. This matrix contains the evaluation function $f_i(L_{j,i}, g_k)$ of each learner $L_{j,i} \in A'_i$ on each criterion $g_k \in F_1$ such that $i \in \{1..4\}$, $j \in \{1..30\}$ and $k \in \{1..11\}$. Then, based on its expertise and the information table, the pedagogical team assign each of the "Learners of reference" in one of the three decision classes.

Phase 2: The Inference of a Preference Model Based on the Incremental Update of the DRSA Approximations. At the end of each week W_i during the MOOC broadcast such that $i \in \{1..4\}$, we applied the algorithm *DRSA-Incremental* on the decision table built at the same week W_i in order to update the approximations of the unions of all the decision classes. Then, the updated approximations were provided as an input to the algorithm *DOMLEM* [11] in order to infer the decision rules.

Phase 3: The Classification of the "Potential learners". At the beginning of each week W_i of the MOOC; such that $i \in \{2..5\}$ we used the inferred decision rules in order to assign the "Potential learners" in one of the three decision classes: Cl_1 of the "At-risk learners", Cl_2 of the "Struggling learners" or Cl_3 of the "Leader learners". The "Potential learners" are those who filled their registration form and, therefore, they were likely to be classified in one of the three decision classes.

6.3   Results and Discussion

This section presents the results given when applying the preference model on real data coming from the French MOOC.

All the algorithms in this paper are coded by Java and run on a personal computer with Windows 7, Intel (R) $Core^{TM}$ i3-3110M CPU @ 2.4 GHz and 4.0 GB memory.

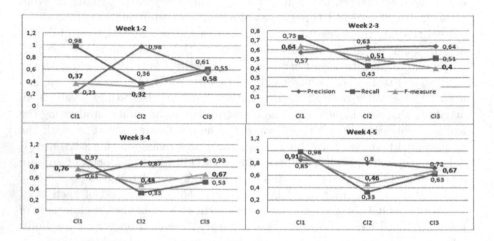

Fig. 3. The efficiency measures of the prediction model for each decision class over weeks

Figure 3 presents the precision, the recall and the F-measure measures relative to each decision class Cl_1, Cl_2 and Cl_3 for each week during the broadcast of the concerned MOOC. Bellow, we consider only the F-measure since it is the harmonic measure between the recall and the precision ones.

- The F-measure corresponding to the Cl_1 decision class of the "At-risk learners" increases over time. Thus, the efficiency of the Cl_1 class prediction increases from a week to another. In effect, a MOOC is known by the presence of what we call "lurkers". These are the participants who register just to discover the MOOC concept and who leave it at the first evaluation. And in spite of their activity during the first week of the MOOC, they keep having the prior intention to abandon it. This type of learners degrades the quality of the prediction model which is based on the profile and the behaviour of the learner and not on his intention. Consequently, the fewer the number of lurkers gets, the higher the prediction quality becomes.
- The F-measure relative to the Cl_3 decision class of the "Leader learners" increases over time too. In fact, from one week to another, these learners enhance their participation in the forum, a thing which gives us more information concerning their profile. In addition, the assessment activities provided by the MOOC are increasingly complex over the weeks. Obviously, if

compared to a simple Quiz, a complex assessment such as the peer-to-peer activity permits more to better assess a learner. This is justified by the deterioration of the F-measure of the Cl_3 class in the week 2–3. In fact, the MOOC on which we tested this approach, proposed a quiz at the end of week 2 and a peer to peer activity at the end of week 3. However, students who pass a quiz may hang at a peer-to-peer activity or even abandon the MOOC because of its complexity. This may also affect the prediction quality.

To sum up, we can confirm that, based on these results, the DRSA approach has achieved very satisfactory results (overall F-measure reached 0.66 for the week 4–5 and 0.61 for the week 1–2). Moreover, the incremental approach allowed to have a richer and a more diverse sample.

7   Conclusion

In this paper, we proposed a multicriteria classification approach based DRSA, called *MAI2P*. It aims at the incremental periodic prediction and consists of three phases: the construction of a decision table, the inference and the update of a preference model that results in a set of decision rules, and the assignment of each potential action in one of the predefined decision classes. The first phase is based on three steps: (i) the construction of a representative training sample called "Actions of reference", (ii) the construction of a coherent family of criteria for the actions characterization, and (iii) construction of the decision table. The approach *MAI2P* depends on the human decision maker preferences.

Validated in a MOOC context, experiments showed that the proposed approach gives an efficient preference model with an F-measure that reaches 0.66. Our perspectives aim to validate the *MAI2P* approach in other fields and to compare it with other conventional machine learning techniques.

References

1. Balakrishnan, G.: Predicting student retention in massive open online courses using hidden Markov models. Technical report No. UCB/EECS-2013-109, University of California at Berkeley (2014)
2. Barak, M.: The same MOOC delivered in two languages: examining knowledge construction and motivation to learn. In: Proceedings of the EMOOCS, pp. 217–223 (2015)
3. Beccali, M., Cellura, M., Mistretta, M.: Decision-making in energy planning application of the electre method at regional level for the diffusion of renewable energy technology. Renew. Energy **28**(13), 2063–2087 (2003)
4. Bouzayane, S., Saad, I.: A preference ordered classification to leader learners identification in a MOOC. J. Decis. Syst. **26**(2), 189–202 (2017)
5. Bouzayane, S., Saad, I.: Incremental updating algorithm of the approximations in DRSA to deal with the dynamic information systems of MOOCs. In: International Conference on Knowledge Management, Information and Knowledge Systems, pp. 55–66 (2017)

6. Chaplot, D., Rhim, E., Kim, J.: Predicting student attrition in MOOCs using sentiment analysis and neural networks. In: Proceedings of AIED 2015 Fourth Workshop on Intelligent Support for Learning in Groups (2015)
7. Chaturvedi, S., Goldwasser, D., Daume II, H.: Predicting instructor's intervention in MOOC forums. In: Proceedings of the 52nd Annual Meeting of the Association for Computational Linguistics (ACL), pp. 1501–1511 (2014)
8. Chevalier, J., Rousseaux, P.: Classification in LCA: building of a coherent family of criteria. Int. J. Life Cycle Assess. 4(6), 352–356 (1999)
9. Delen, D., Walker, G., Kadam, A.: Predicting breast cancer survivability: a comparison of three data mining methods. Artif. Intell. Med. 34(2), 113–127 (2005)
10. Fini, A.: The technological dimension of a massive open online course: the case of the CCK08 course tools. Int. Rev. Res. Open Distance Learn. 10(5), 1–26 (2009)
11. Greco, S., Matarazzo, S., Slowinski, S.: Rough sets theory for multicriteria decision analysis. Eur. J. Oper. Res. 129, 1–47 (2001)
12. Michalski, R.S., Reinke, R.E.: Incremental learning of concept descriptions: a method and experimental results. In: Machine Intelligence XI, pp. 263–288. Clarendon Press, Oxford (1988)
13. Miller, G.: The magical number seven, plus or minus two: some limits on our capacity for processing information. Psychol. Rev. 63, 81–97 (1956)
14. Moon, S., Potdar, S., Martin, L.: Identifying student leaders from MOOC discussion forums. In: Conference on Empirical Methods in Natural Language Processing, pp. 15–20 (2014)
15. Morris, N., Hotchkiss, S., Swinnerton, B.: Can demographic information predict MOOC learner outcomes? In: Proceedings of the EMOOC Stakeholder Summit, pp. 199–207 (2015)
16. Roy, B., Mousseau, V.: A theoretical framework for analysing the notion of relative importance of criteria. J. Multi-criteria Decis. 5, 145–159 (1996)
17. Pawlak, Z.: Rough sets. Int. J. Comput. Sci. 11, 341–356 (1992)
18. Saad, I.: Une contribution mthodologique pour l'aide l'identification et l'évaluation des connaissances nécessitant une opération de capitalisation. Thèse de doctorat, Université Paris-Dauphine (2005)
19. Sarukkai, R.R.: Link prediction and path analysis using Markov chains. Comput. Netw. 33(1), 377–386 (2000)
20. Shanahan, M.: Prediction is deduction but explanation is abduction. In: International Joint Conference on Artificial Intelligence, pp. 1055–1060 (1989)
21. Sharma, A., Sahoo, P.K., Tripathi, R., Meher, L.C.: Artificial neural network-based prediction of performance and emission characteristics of ci engine using polanga as a biodiesel. Int. J. Ambient Energy 37(6), 559–570 (2016)
22. Suriel, S.L., Atwater, M.M.: From the contributions to the action approach: white teacher. J. Res. Sci. Teach. 49(10), 1271–1295 (2012)
23. Wolff, A., Zdrahal, Z., Herrmannov, D., Kuzilek, J., Hlosta, M.: Developing predictive models for early detection of at-risk students on distance learning modules. In: Workshop on Machine Learning and Learning Analytics at LAK (2014)
24. Xing, W., Chen, X., Stein, J., Marcinkowski, M.: Temporal predication of dropouts in MOOCs: reaching the low hanging fruit through stacking generalization. Comput. Hum. Behav. 58, 119–129 (2016)
25. Yang, D., Sinha, T., Adamson, D., Rose, C.: Turn on, tune in, drop out: anticipating student dropouts in massive open online courses. In: NIPS Workshop on Data Driven Education (2013)

Assessing Multidimensional Sustainability of European Countries with a Novel, Two-Stage DEA

Georgios Tsaples[1]([⊠]), Jason Papathanasiou[1], Andreas C. Georgiou[1], and Nikolaos Samaras[2]

[1] Department of Business Administration, University of Macedonia, Egnatia Str. 156, 54636 Thessaloniki, Greece
gtsaples@gmail.com, {jasonp,acg}@uom.edu.gr
[2] Department of Applied Informatics, University of Macedonia, Egnatia Str. 156, 54636 Thessaloniki, Greece
samaras@uom.edu.gr

Abstract. The issues of sustainability and sustainable development are considered important aspects of governmental policy making. Currently, many methods are used to assess the sustainability of various regions and countries each accompanied by advantages and disadvantages. The objective of the current paper is to propose a new methodological framework for the assessment of sustainability that attempts to mitigate some of the limitations of the methods that are used. The proposed method is based on two-stage Data Envelopment analysis. In the first stage, raw data are transformed to sub-indicators using the multiplicative version of the VRS DEA model. The sub-indicators are used in the second-stage, in a typical Benefit-of-the-Doubt model, to calculate their optimal weights, which are used in the construction of a geometric composite indicator of sustainability. The method is tested to calculate a sustainability index for 28 European countries. The results show that eastern European and Scandinavian countries appear to be more sustainable than western, developed countries.

Keywords: Sustainability · Data Envelopment Analysis · Benefit-of-the-Doubt · Composite indicator

1 Introduction

Sustainable development in its most broad sense means the ability to address the needs of the present without compromising the ability of future generations to do the same for their own needs [1]. Furthermore, sustainability, as a measure of sustainable development, is a multi-dimensional concept [2]; it is a structure that includes economic, environmental and social dimensions [3–6].

A series of international treaties along with the Sustainable Development Goals of the European Union have demonstrated how sustainability has been increasingly the focus of governmental activity [7]. However, not much is offered in terms of methodological guidance or tools as to how to achieve sustainable development. Such a

© Springer Nature Switzerland AG 2019
P. S. A. Freitas et al. (Eds.): EmC-ICDSST 2019, LNBIP 348, pp. 111–122, 2019.
https://doi.org/10.1007/978-3-030-18819-1_9

guidance could assist public and private organizations in their effort to assess their state of sustainable development.

The use of appropriate methods to address this objective can assist policy makers in reaching appropriate decisions [8]. Especially, since complexity and uncertainty have become an inherent characteristic of modern-day policy making. It is this complexity that renders methods that rely on cost-benefit analyses limiting to fully capture the issue of sustainability [9].

Thus, the objective of the present paper is to propose a new method to measure sustainability that is based on Data Envelopment Analysis. To test the suitability and appropriateness of the new method, the sustainability of EU countries is measured as a case study.

The rest of the paper is organized as follows: Sect. 2 is focused on a literature review on DEA and sustainability. Gaps are identified in this section. In Sect. 3 the proposed method is established while the case study is presented in Sect. 4. Conclusions and future research directions are discussed in the last section of the paper.

2 Literature Review

Data Envelopment Analysis (DEA) is a non-parametric method that is used for the assessment of the technical efficiency of Decision Making Units (DMUs) relative to one another [10, 11]. In DEA, technical efficiency is a measure of how well a Decision Making Unit can transform inputs into outputs. The foundations of DEA can be traced in the works of [12–14] and the original development of the method was presented in the seminal papers of [15] and [16].

DEA uses a mathematical programming approach for the evaluation of the performance of DMUs where price information may not be available [17]. Furthermore, it does not require knowledge on the type of relationships between inputs and outputs, there are no pre-requisites regarding the statistical distribution of data and finally, it can provide insights into how to improve the performance of an inefficient DMU [8].

These advantages were crucial in identifying that the method could be suitable for aggregating economic, environmental and social indicators, hence proving to be appropriate in a sustainability context [18].

Zhou et al. [19] performed an extensive literature review on the use of DEA and sustainability. Some of the most important patterns that the authors detected in the literature are: DEA has been increasingly used to assess sustainability, as identified by the literature review of [19]. In more detail, studying the literature it can be observed that early adopters of the method used mainly the classical DEA models, while as their familiarity with the method increases so does the development of methodological variations. Finally, a trend is observed, where different DEA variations and/or other methodologies are combined in the context of sustainability.

However, the authors identified also several gaps in the literature. Firstly, the main focus of the applications is on the study of economic and environmental measures. The measure of "eco-efficiency" is increasingly used, however, the integration of the social dimension is lacking. Moreover, there is an ongoing debate on which indicators and data should be used to represent the various dimensions of sustainability, while finally, the majority of areas that the method was used is in China and its regions. Other areas of the world, including EU countries have not been studied as extensively. Hence, it can be concluded that applications focusing in European regions are still low in numbers and value could be gained by continuously assessing how they are faring in

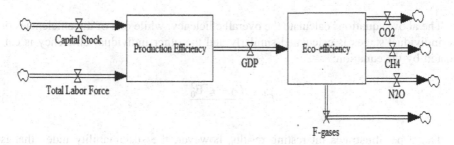

Fig. 1. The two-stage model of [20]

achieving the Sustainable Goals.

One of the most interesting uses of DEA for the assessment of sustainability was performed in the paper of Halkos et al. [20]. In their article the authors used an additive two-stage DEA to calculate a sustainability indicator for 20 countries of advanced economies. Their two-stage model is illustrated in the figure (Fig. 1) below.

The overall efficiency is calculated from the efficiencies of each step as:

$$E_0 = \xi_1 * \frac{\sum_{d=1}^{D} w_d z_{d0} + u^A}{\sum_{i=1}^{m} u_i x_{i0}} + \xi_2 * \frac{\sum_{r=1}^{s} u_r y_{r0} + u^B}{\sum_{d=1}^{D} w_d z_{d0}} \tag{1}$$

ξ_1, ξ_2 *indicate the weight for each stage*
x_{ij} *indicates the input i of DMU j of stage 1*
z_{dj} *indicates the output d of DMU j of stage 1 or the input of stage 2*
y_{rj} *indicates the output r of DMU j of stage 2*
u_i, w_d, u_r *are the weights for each input and output of each stage*

The additive two-stage model is calculated under Variable Returns to Scale as follows:

$$\max \sum_{d=1}^{D} \mu_d z_{d0} + \sum_{r=1}^{s} \gamma_r y_{r0} + u^1 + u^2 \tag{2}$$

$$s.t. \sum_{i=1}^{m} \omega_i x_{i0} + \sum_{d=1}^{D} \mu_d z_{d0} = 1 \tag{3}$$

$$\sum_{d=1}^{D} \mu_d z_{dj} - \sum_{i=1}^{m} \omega_i x_{ij} + u^1 \leq 0, \ j = 1,2,\ldots n \tag{4}$$

$$\sum_{r=1}^{s} \gamma_r y_{rj} - \sum_{d=1}^{D} \mu_d z_{dj} + u^2 \leq 0, \ j = 1,2,\ldots n \tag{5}$$

$$\gamma_r, \mu_d, \omega_i \geq 0, \ \textit{the optimal multipliers} \tag{6}$$

$$i = 1\ldots m, \ j = 1\ldots n, \ r = 1\ldots s, \ d = 1\ldots D, \ u^1, \ u^2 \textit{ free} \tag{7}$$

The above equations calculate the overall efficiency, while one of the efficiencies of the individual stages is calculated similarly and the other individual efficiency is calculated by the equation:

$$E_0^1 = \frac{E_0 - \xi_2^* E_0^2}{\xi_1^*} \tag{8}$$

The paper illustrates interesting results, however, the sustainability index that is calculated lacks a social dimension as it relies heavily on economic and environmental figures.

Another stream of research within DEA is the construction of composite indicators using the logic and assumptions of the method. Composite indicators have gained attraction because they are easy to communicate and they have the potential to measure multi-dimensional concepts that may not share common units of measurement [21].

The use of DEA for the construction of composite indicators has increased in many areas of applications. One of the more recent methodological variations of the method was developed by [22]. The authors choose a specific set of base performance indicators y_{rB} and for each DMU the following model is solved:

$$\max \sum_{r=1}^{s} w_{ri} y_{ri} \tag{9}$$

$$s.t. \sum_{r=1}^{s} w_{ri} y_{rj} \leq 1 (N \textit{ constraints, one for each DMU } j = 1\ldots N) \tag{10}$$

$$w_{rj} \geq 0, \ (s \textit{ constraints one for each sub}-\textit{indicator}) \tag{11}$$

The linear model of Eqs. (9)–(11) is the typical Benefit-of-the-Doubt model (based on the principles of DEA). The authors use it to calculate the optimal weights for the construction of the final indicator.

However, the composite indicator is calculated by the equation:

$$CI_j = \prod_{r=1}^{s} \left(\frac{y_{rj}}{y_{rb}}\right)^{\omega_r^*} \tag{12}$$

Hence, the indicator is a geometric one. Furthermore, the values for y_{rB} are considered base values on which the numerators is compared, thus keeping the composite indicator unit invariant. Furthermore, the values of ω_r^* indicate how much the rth subindicator contributes to the final outcome.

The authors use their method to calculate social inclusion in EU countries. However, one limitation of the approach is similar to a limitation of the DEA methodology in general; there is a necessity to make a trade-off between the number of inputs and outputs in order to avoid situations where all the DMUs are considered efficient. As a result, the methodological limitations that may have affected the sub-indicators have been transferred to the construction of the final composite index, despite the fact that the use of a geometric composite indicator, mitigates the disadvantages of using an additive approach.

3 Model Formulation

The proposed model attempts to address the methodological gaps that were identified in the previous section and it is based on the works of [20] and [22].

More analytically, the proposed model will be a combination of the two approaches. Starting from raw data, each dimension of sustainability will be calculated using the two-stage approach by Halkos et al. [20], once for each of the economic, environmental and social dimensions, following the Eqs. (2)–(7).

Consequently, the three sub-indicators will be used as the input for the model by [22], following the Eqs. (9)–(12), where the sub-indicators of the first stage are used. The result will be a geometric, composite indicator that is multilayered and multidimensional. Multilayered because for its construction the limitation identified in the Benefit-of-the-Doubt models regarding the number of inputs, outputs and the availability of raw data, is addressed in the first stage of the model, where each dimension is calculated separately. Multidimensional because each of the dimensions of sustainability will be equally represented in the final composite indicator.

3.1 Choice of Data and Indicators

For the case study we are going to start by a dimension as calculated by Halkos et al. [20]. Their indicator relies on the notion of eco-efficiency that can be considered as a measure of economic and environmental performance. However, as their indicator does not include a social dimension, it cannot be considered an accurate measure of sustainability. Furthermore, sustainability implies constant research and innovation in order to achieve technologies that mitigate the harmful effects to the environment. The particular definition is interpreted from a report from the European Commission [23] that justifies that growth should be:

- Smart: since knowledge and innovation will drive the current and future economy
- Sustainable: meaning that is resource-saving, energy efficient etc.
- Inclusive: meaning that it involves and represents large parts of the population. For the particular case study, inclusive is represented by factors that illustrate the satisfaction of citizens for their respective countries. As a result, inside the definition

As a result, it can be observed that already for the construction of the sustainability index, more than one perceptions are included: those of the authors, the EU's, citizens' etc. The notion of sustainability has proven difficult to define and consequently remains a contested one. By incorporating more than one perceptions, it is ensured that limitations that (and objections) that step from its definition could be mitigated.

For the current case study, three dimensions are considered. The variables that are considered serve only as an example for the illustration of the proposed method. A more comprehensive and detailed choice will be deployed in future versions of the paper. These are:

- Economic-Environmental
  - Inputs: Gross Fixed Capital in PPS, Total Labor Force
  - Outputs/Second stage input: GDP per capita in PPS
  - Outputs: Share of renewable energy in gross final energy consumption, Greenhouse gas emissions (in CO2 equivalent)
- Economic-Social
  - Inputs: Gross Fixed Capital in PPS, Total Labor Force
  - Outputs/Second stage input: GDP per capita in PPS
  - Outputs: Overall life satisfaction, Satisfaction with living environment, Satisfaction with financial situation
- Economic-Research and Development
  - Inputs: Gross Fixed Capital in PPS, Total Labor Force
  - Outputs/Second stage input: GDP per capita in PPS
  - Outputs: Intramural R&D expenditure for all sectors of the economy

The table below illustrates the above variables and the values that they get, as taken from [24]. The data are for 28 European Countries, the ones that are most often used in the statistical information provided by the EU (Table 1).

Table 1. Data and variables for the 28 European countries that are used in the model

Country	Economic-Environmental					Economic-Social				Economic Research and Development
	Gross fixed capital at current prices (PPS)	Total Labour force (×1000 persons)	GDP per capita in PPS-Index (EU28 = 100)	Share of renewable energy in gross final energy consumption	Greenhouse gas emissions (in CO_2 equivalent)	Overall life satisfaction	Satisfaction with living environment	Satisfaction with financial situation	Intramural R&D expenditure (GERD) by sectors of performance [Euro per inhabitant]	
Belgium	79.2	5035.9	120	7.5	82.5	7.6	7.6	6.9	822.1	
Bulgaria	18.8	3857.6	46	19	53.5	4.8	5.2	3.7	36.6	
Czech Republic	59	5406.6	84	13.8	65	6.9	7.5	6.0	285	
Denmark	36.7	2934	128	27.4	79.9	8.0	8.2	7.6	1,371.8	
Germany	527.9	44439	124	12.4	76.6	7.3	7.7	6.3	990.1	
Estonia	7.4	683.1	75	25.6	54.2	6.5	6.8	5.4	247	
Ireland	30.4	2198.6	132	7.7	105.4	7.4	8.0	5.5	610.3	
Greece	25.6	5327.7	72	15	99.4	6.2	6.2	4.3	133.2	
Spain	209.4	23962	89	15.3	114.6	6.9	7.2	5.8	278.5	
France	422.4	30612	108	14.1	90.2	7.1	7.6	6.4	722	
Croatia	13.4	1854	60	28	77	6.3	6.3	4.6	83.2	
Italy	275.7	26645.9	98	16.7	86.1	6.7	6.0	5.7	351.6	
Cyprus	2.7	440.3	84	8.1	137.2	6.2	6.0	5.2	101	
Latvia	7.8	1017.3	62	37.1	43.8	6.5	7.2	5.0	69.1	
Lithuania	10.7	1465.2	73	22.7	41.6	6.7	7.8	5.8	111.9	
Luxemburg	7.4	249.2	261	3.5	93.6	7.5	7.8	6.9	1,127.9	
Hungary	37.1	4333.8	67	16.2	61.3	6.1	6.5	5.2	142.8	
Malta	1.7	193.3	85	3.7	139.4	7.1	7.1	6.0	139.8	
Netherlands	111.6	9248	134	4.8	90.7	7.8	8.0	7.4	759.6	
Austria	68.8	4387.1	131	32.4	103.3	7.8	8.4	7.0	1,132.4	
Poland	129.9	17361	67	11.4	84.7	7.3	7.6	5.7	90.3	
Portugal	31.6	5339.8	77	25.7	110.9	6.2	6.3	4.5	215.4	
Romania	72	9376.6	54	23.9	46.8	7.1	7.4	6.2	27.9	
Slovenia	8.9	1031.4	82	22.4	98.9	7.0	7.7	5.6	454.1	
Slovakia	23	2715.3	77	10.1	57.7	7.0	6.9	5.5	112.9	
Finland	35	2738.6	113	36.7	89.9	8.0	7.8	7.5	1,231.7	
Sweden	72.3	5115.7	125	52	79.2	7.9	7.6	7.5	1,507.6	
United Kingdom	297.3	32749	108	5.7	73.8	7.3	7.8	6.2	532	

The figure below (Fig. 2) illustrates the overall process for the construction of the sustainability composite indicator.

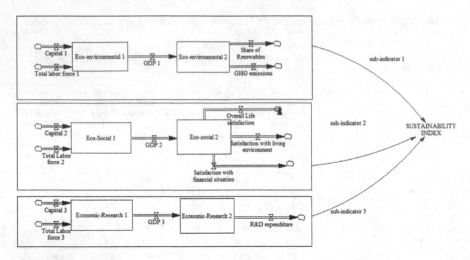

Fig. 2. The process for the construction of the composite sustainability indicator

4   Results

The data for the case study were taken from [24]. The first part of the method is focused on the construction of the sub-indicators as were defined in the previous sections. The table (Table 2) and the figure below (Fig. 3) illustrate and plot in the European Map the three sub-indicators that were calculated in the first stage of the method.

Table 2. Results for the sub-indicators and the composite sustainability indicator

Country	Econ-Environmental	Econ-Social	Economic R&D	Sustainability Composite Indicator
Belgium	0.2115	0.3815	0.387	0.772644442
Bulgaria	0.545	0.545	0.545	1.271245497
Czech Republic	0.29135	0.367	0.37085	0.798479541
Denmark	0.25075	0.5421	0.50185	1.03249499
Germany	0.213	0.3355	0.4195	0.765779832
Estonia	0.487	0.4874	0.52295	1.163457827
Ireland	0.22695	0.4199	0.34345	0.779467143
Greece	0.3522	0.3862	0.3882	0.873790001
Spain	0.262365	0.3033	0.33536	0.699283414
France	0.2157765	0.3272	0.38627	0.730201513
Croatia	0.5125	0.1056	0.467	1.009414488
Italy	0.2383	0.2718	0.3248	0.649406623
Cyprus	0.588315	0.6173	0.60881	1.4088848
Latvia	0.6089745	0.5384	0.49397	1.288936232

(continued)

Table 2. (*continued*)

Country	Econ-Environmental	Econ-Social	Economic R&D	Sustainability Composite Indicator
Lithuania	0.579439	0.5794	0.42193	1.248237598
Luxemburg	0.588	0.674	0.7	1.521161277
Hungary	0.365911	0.4019	0.40791	0.911837239
Malta	0.7705	0.8175	0.803	1.856131183
Netherlands	0.186	0.4435	0.331	0.771811714
Austria	0.244	0.5235	0.4235	0.947591245
Poland	0.3495	0.5065	0.371	0.959210002
Portugal	0.3635	0.3565	0.3875	0.861022856
Romania	0.511805	0.5118	0.43680	1.141568251
Slovenia	0.394	0.522	0.5125	1.108561659
Slovakia	0.33545	0.3849	0.36195	0.839511446
Finland	0.3097	0.5372	0.5247	1.077038745
Sweden	0.5207	0.4727	0.5207	1.180680284
United Kingdom	0.21661	0.384	0.34011	0.738939253

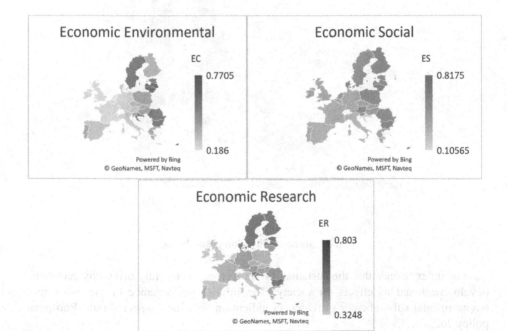

Fig. 3. The sub-indicators from the first stage of the method

Several insights can be gained already from the first-stage results. It appears that under the particular definitions for the sub-indicators (with the particular inputs and outputs), countries of East Europe seem to fare better than those from the traditional west. These countries became members of the union later in the EU's life. As a result, it appears that their current economic development has not created the problems that seem to burden the western, "developed" countries. However, observing the sub-indicators, another insight is revealed: in the majority of the cases the economic-environmental sub-indicator appears to fare worse than the other two. As a result, the sustainability of these countries is driven more by their economic growth and its consequences on their society and less on their attention to the environment.

Only Luxemburg, Cyprus, Malta and the Scandinavian countries appear to follow the trend of the eastern countries, as it performs well in all the sub-indicators. However, similar to the previous countries, the efficiency with regards to the environment appears to perform worse than in the other two dimensions.

The figure below (Fig. 4) illustrates the overall sustainability composite indicator, as constructed in the second stage of the method.

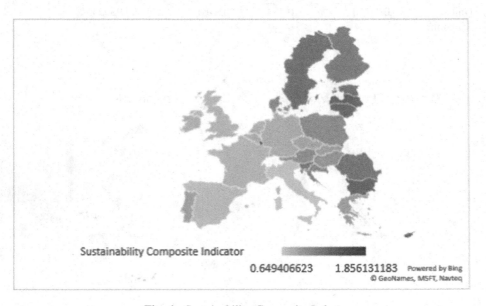

Fig. 4. Sustainability Composite Index

The index reveals that the sustainability in Europe is currently driven by economic development and its effects on society. The inferior performance of the economic-environmental sub-indicator provides an indication on which aspect should European policy focus.

Finally, it should be mentioned that the mathematic formulation that was used is considered "optimistic" since it calculates the importance weights favoring the most advantageous sub-indicators. As a result, the trend that characterized the sub-indicators continues into the sustainability index. To increase the robustness of the results, validation processes will be employed in future directions of the paper, with the purpose of formally identifying the robustness of the model and taking measures that stabilize its performance.

5 Conclusions and Future Research

The purpose of this paper was to propose a two-stage Data Envelopment Analysis variation for the construction of a multi-dimensional sustainability index. In the first stage of the method, raw data are transformed into indicators that represent three dimensions of sustainability. They are focused on the performance of the economy and the environment, the economy and society and the economy and research/innovation.

These sub-indicators are used in a typical Benefit-of-the-Doubt DEA model with the purpose of obtaining their optimal weights. Subsequently, they are aggregated geometrically for the construction of the overall sustainability composite index.

The combination of the two approaches into one methodological framework, mitigates some of the limitations that each approach had when used standalone. As a case study, data were used for 28 European countries and the results demonstrated that traditional (western) countries fare comparatively worse than those that came later in the Union.

However, the method is not without its limitations. For the construction of the sub-indicators, different approaches could be followed: different combinations of inputs and outputs or different thematic contexts. For example, the eco-social dimension could focus more heavily on employment or social welfare that do not appear in the current model. As a result, there is the need to consider different interpretations of the dimensions of sustainability and sustainability itself.

However, when different interpretations are considered, there is the need to aggregate them into one meaningful index. Furthermore, there is the need to incorporate in such measurements, an indication of reliability. For example, different weights could be attributed to a definition of sustainability by someone who has worked on the subject than someone who has not and possesses only a passing knowledge of the issue. Both the perceptions should be considered, but with different credibility.

Finally, the inherent mathematical nature of the method can make it difficult in its use by non-specialists/experts. As a result, there is the need for a comprehensive and user-friendly interface that would allow policy makers and citizens to experiment with the measurement of performance. At the same time, the interface could gain extra value if it was accompanied by automated explanations. Hence, the results would not be limited to a simple number/value, but could communicate what the particular value means and how it was produced.

All these limitations constitute future directions of research for the proposed methodology.

References

1. Brundtland, G.: Report of the World Commission on environment and development: our common future, United Nations (1987)
2. Munda, G., Saisana, M.: Methodological considerations on regional sustainability assessment based on multicriteria and sensitivity analysis. Reg. Stud. **45**(2), 261–276 (2011)

3. Robinson, J.: Squaring the circle? Some thoughts on the idea of sustainable development. Ecol. Econ. **48**(4), 369–384 (2004)
4. Snedoon, C., Howarth, R., Norgaard, R.: Sustainable development in a post-Brundtland world. Ecol. Econ. **57**(2), 253–268 (2006)
5. Holden, E., Linnerud, K., Banister, D.: Sustainable development: our common future revisited. Glob. Environ. Change **26**, 130–139 (2014)
6. Klumpp, M.: Do forwarders improve sustainable efficiency? Evidence from a European DEA Malmquist index calculation. Sustainability **9**(5), 842 (2017)
7. Coli, M., Nissi, E., Rapposelli, A.: Monitoring environmental efficiency: an application to Italian provinces. Environ. Model Softw. **26**(1), 38–43 (2011)
8. He, J., Wan, Y., Feng, L., Ai, J., Wang, Y.: An integrated data envelopment analysis and emergy-based footprint methodology in evaluating sustainable development, a case study of Juangsu province, China. Ecol. Ind. **70**, 23–34 (2016)
9. Adler, M.: Well-Being and Fair Distribution: Beyond Cost-Benefit Analysis. Oxford University Press, Oxford (2012)
10. Thanassoulis, E.: Introduction to the Theory and Application of Data Envelopment Analysis. Kluwer Academic Publishers, Dordrecht (2001)
11. Førsund, F.R., Sarafoglou, N.: On the origins of data envelopment analysis. J. Prod. Anal. **17**(1–2), 23–40 (2002)
12. Debreu, G.: The coefficient of resource utilization. Econometrica: J. Econ. Soc. **19**, 273–292 (1951)
13. Farrell, M.: The measurement of productive efficiency. J. R. Stat. Soc. Ser. A **120**(3), 253–290 (1957)
14. Diewert, W.: Functional forms for profit and transformation functions. J. Econ. Theory **6**(3), 284–316 (1973)
15. Charnes, A., Cooper, W., Rhodes, E.: Measuring the efficiency of decision-making units. Eur. J. Oper. Res. **3**(4), 429–444 (1978)
16. Banker, R., Charnes, A., Cooper, W.: Some models for estimating technical and scale inefficiencies in data envelopment analysis. Manag. Sci. **30**(9), 1078–1092 (1984)
17. Kuosmanen, T., Kortelainen, M.: Measuring eco-efficiency of production with data envelopment analysis. J. Ind. Ecol. **9**(4), 59–72 (2005)
18. Callens, I., Tyteca, D.: Towards indicators of sustainable development for firms: a productive efficiency perspective. Ecol. Econ. **28**(1), 41–53 (1999)
19. Zhou, H., Yang, Y., Chen, Y., Zhu, J.: Data envelopment analysis application in sustainability: the origins, development and future directions. Eur. J. Oper. Res. **264**(1), 1–16 (2018)
20. Halkos, G., Tzeremes, N., Kourtzidis, S.: Measuring sustainability efficiency using a two-stage data envelopment analysis approach. J. Ind. Ecol. **20**(5), 1159–1175 (2016)
21. OECD: Handbook on Constructing Composite Indicators: Methodology and User Guide. OECD, Paris (2008)
22. Van Puyenbroeck, T., Rogge, N.: Geometric mean quantity index numbers with Benefit-of-the-Doubt weights. Eur. J. Oper. Res. **256**(3), 1004–1014 (2017)
23. EC (European Commission): Europe 2020: A European Strategy for Smart, Sustainable and Inclusive Growth. European Commission, Brussels (2010)
24. EUROSTAT: Database. Eurostat (2018)

Logikós: Augmenting the Web with Multi-criteria Decision Support

Alejandro Fernández[1](✉)(ID), Gabriela Bosetti[1](ID), Sergio Firmenich[1](ID),
and Pascale Zaraté[2](ID)

[1] LIFIA, CIC/Facultad de Informática, UNLP,
50th Street and 120 Street, 1900 La Plata, Argentina
{alejandro.fernandez,gabriela.bosetti,
sergio.firmenich}@lifia.info.unlp.edu.ar
[2] IRIT, Toulouse Université, 2 rue du Doyen Gabriel Marty,
31042 Toulouse Cedex 9, France
pascale.zarate@irit.fr

Abstract. There are activities that on-line customers daily perform,
which involve a multi-criteria decision challenge. Choosing a destination
for traveling, buying a book to read, or buying a mobile phone are some
examples. Customers analyze and compare alternatives considering a set
of shared characteristics, and under certain criteria. E-commerce websites frequently present the information of products without special support to compare them by one or many properties. Moreover support for
decision making is limited to sorting, filtering, and side-by-side comparison tables. Consequently, customers may have the feeling that the merchants interests influence their choices, which are no longer grounded on
the rational arguments they would like to put in practice. Moreover, the
alternatives of interest for the customer are frequently scattered across
various shops, with no support to collect and compare them in a consistent and customized manner. In this article, we propose empowering
users with multi-criteria decision making support on any website, and
across different websites. We also present Logikós, a toolbox supporting
multi-criteria decision making depending on the presentation layer of any
Web page.

Keywords: E-commerce · Multi-criteria decision support ·
Web augmentation

1 Introduction

Buying online has become a widespread and increasing practice in the world.
According to a report on July 2017 by Statista[1], the number of worldwide online
buyers is expected to increase from 1.32 billion in 2014 to 2.14 billion in 2021,
with no decreases in the middle years.

---

[1] https://www.statista.com/statistics/251666/number-of-digital-buyers-worldwide/.

© Springer Nature Switzerland AG 2019
P. S. A. Freitas et al. (Eds.): EmC-ICDSST 2019, LNBIP 348, pp. 123–135, 2019.
https://doi.org/10.1007/978-3-030-18819-1_10

Buying online can be seen as a multi-criteria decision challenge since it usually involves comparing not one but many properties of a same kind of product (evaluating different alternatives), under the same criteria to make a decision using a concrete process. For instance, Akarte et al. [1] identified 18 criteria for casting supplier assessment (as sample delivery time, maximum casting size or minimum section thickness).

A problem with today's e-commerce websites is that they do not help users to make decisions in a personalized and user-controllable way. There are e-commence aids to compare products, but these are usually limited to ranking, filtering, and side-by-side property tables. In this sense, e-commerce seems to ignore years of development in decision support systems, which successfully offer straightforward criteria aggregation operators such as the Weighted sum and even richer ones such as the "Ordered weighted sum", "Choquet Integral". After using existing e-commerce platforms, the customers may be left with the feeling that the merchants' interests influence their choices, which are no longer grounded on rational arguments.

Another problem is that the alternatives the customer wants to consider might not all be available in a single e-commerce website, or that the information composing a specific alternative might be available across different sites and with different prices. Customers are not very well supported to collect and compare the alternatives available in various sites in a consistent manner. A solution to this challenge is available in the form of meta-search engines but only for some domains, as Trivago[2] for hotels and flights.

In this article, we propose an approach to empower users with multi-criteria decision making support on any website, and across different websites. We illustrate our approach with a descriptive evaluation in the domain of agricultural machinery. We also present Logikós, a toolbox supporting such decision making approach.

The article is organized as follows. Section 2 introduces key background concepts regarding decision support systems and multi-criteria decision making. Section 3 presents the approach. A descriptive evaluation in the form of a scenario is presented in Sect. 4. Related Work and the conclusions follow.

2 Background

2.1 Informed Decision Making

The information and the tools we might have to support purchase decisions are disconnected. Even though the Web is full of information that we could use to make rational purchase decisions, we hardly use it. Information is ill-structured and incomplete. Information is hard to find. Making a decision based on information sometimes requires complex (e.g., multi-criteria) decision processes, which many people do not know.

---

[2] http://www.trivago.com Last accessed December 2018.

When the circumstances make it worthwhile, some decision-makers resort to spreadsheets [6]. A simple web-search for "spreadsheet car shopping" will return multiple hits referring the reader to a variety of prepared spreadsheet templates to help buyers compare and select cars. Such spreadsheets serve two purposes. First, they help users to identify key attributes to look for and record them in a structured manner (although, they still need to harvest the information manually). Second, once attributes have been recorded for all alternatives, the spreadsheet helps to make a decision with calculations and comparisons that reflect the priorities and weights that the template's author gave to each criterion.

We have observed that some students that are proficient in the operation of spreadsheets use them to create similar templates to buy smart-phones, cameras or personal computers. However, in those cases, the use of the spreadsheet is limited to reproducing the comparison functionality commonly offered by e-commerce websites. That is: (a) select candidates, (b) have a general overview of the available properties of all candidates, (c) compare candidates attribute by attribute, (d) sort and filter candidates.

The effort that customers invest in making a decision is directly related to the outcome. There is a trade-off between effort invested and perceived outcome. High stake decisions motivate users to make a more significant effort. When decision makers understand the stakes are not that high, they settle for a sub-optimal alternative, like the first option that meets some threshold criteria or a suggestion from a trusted colleague. This is what the theory of bounded rationality terms "satisficing" [3].

The approach followed by decision makers, like the ones described above, can be characterized by the following process (see Fig. 1). We call it "The spreadsheet strategy for rational purchasing". First, the decision maker describes the items to be purchased (1). The description of the items includes the properties to look for when collecting information. Then, the harvesting starts. The decision maker browses various websites recording information about the items that are a good candidate for purchase (2). The URL of the collected candidates must be recorded so the decision maker can find them again. At some point, the decision maker may identify a new and valuable property, not considered so far. He consequently updates the description of the items and starts collecting the new information from that moment on (1). Ideally, he goes back and updates the information of already collected items that did not include the new properties (2). When an interesting list of candidates is available, the decision maker attempts a decision. For that, he needs to define a decision strategy or use an existing one, like the one created by other users as a spreadsheet template (3), and apply it to rank the so far collected items (4). Applying the decision strategy on already collected items might trigger the need for collecting more items and ever getting rid of some bad performing candidates (2). This iterative process of item modeling, harvesting, and evaluation continues until the optimal candidate is found or the "satisfying" threshold is met.

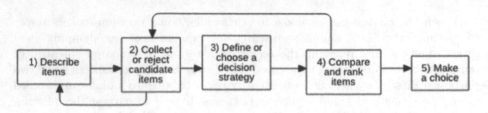

Fig. 1. A model of the spreadsheet strategy for rational purchasing

2.2   Multi-criteria Approach

Several schools of thought widely study Multi-Criteria Decision-Making Analysis (MCDA), and the output of the studied techniques is to obtain a ranked list of alternatives. There are two well-known schools of MCDA: the American school and the European school. The American school is based on an evaluation of a utility function of each alternative. One of the most well-known methods in this school is called the Analytic Hierarchy Process (AHP) that is a mature and widespread process to support decision makers to choose among various alternatives in a context where the decision involves comparing the alternatives according to multiple aspects, each of which contributes differently to the decision. Our approach is based on this method AHP. On the other hand, the European school is not based on the evaluation of a set of matrices but the evaluation of thresholds for each criterion. For each criterion, a preference, indifference and veto thresholds are defined. These thresholds are used to compare alternatives two by two on each criterion and then on several criteria. The two main methodologies are Promethee and Electre. They are less heavy to use than AHP, but the definitions of the thresholds are quite difficult.

AHP helps to identify the aspects to be considered (criteria), supports the exploration of relations among them and their use to choose an alternative. It relies on pairwise comparisons and the judgment of experts to obtain priority scales [8]. AHP has been used a wide variety of domains, as in education, health-care, public administration, telecommunications, manufacturing, and various branches of the Industry (like airlines, defense, entertainment, manufacturing) [7].

At its core, AHP consists of three well-defined steps. First, the decision problem is decomposed into a hierarchy of sub-problems (commonly referred to as decision criteria). Then, decision makers compare the identified sub-problems to each other in order to obtain a model for their relative importance regarding solving the higher level problem. This pairwise comparison among elements in the decision hierarchy is used to obtain the weight or priority of each decision criteria. Such weights represent how much each of the decision criteria contributes to the final decision. At this point, a consistency check can be performed. In any given level, if there are C criteria, $\sum_{n=1}^{C-1} n$ pair-wise comparisons are required.

Russo and Camanho conducted a literature review regarding how criteria is being defined and measured [7]. In a few cases, criteria were obtained from literature on the domain. More cases involved experts in defining/selecting criteria. In this study, we use decision makers in a generic way and the proposed approach is designed for a single user or decision maker. The decision makers preferences are modeled in matrices; if the number of used criteria is n, then the number of matrices to fill in is $n + 1$ depending of the decomposition structure. So, in order to avoid an overload of comparisons, it is important to limit the number of criteria. Literature suggests that the number of criteria and alternatives be kept at seven or less.

Up to this step, the method does not require an analysis of the available alternatives (e.g. products that the customer considers buying). In fact, an interesting aspect of AHP is that it aims to force the decision maker to explicitly reflect on the decision criteria without focusing his attention into the concrete alternatives (thus reducing the preference bias). AHP is about focusing on the factors; it facilitates the decision making by decomposing a decision-problem into factors and organizing them in order to exhibit the forces that influence a decision [8].

In a final step, the alternatives are compared to each other with respect to the decision criteria, which can be tangible (e.g. price) or intangible (e.g. aspirational appeal). If A is the number of available alternatives, and C is the number of criteria, the decision maker needs to make $C * \sum_{n=1}^{A-1} n$ pair-wise comparisons.

Although pairwise comparisons among criteria and among alternatives are not complex per se, the number of required comparisons rises fast. If 4 criteria are considered, and 5 alternatives are available, the number of pair-wise comparisons is computed as $(\sum_{n=1}^{3} n) + 4 * \sum_{n=1}^{4} n = 6 + 4 * 10 = 46$. Moreover, as the number of pairwise comparisons increases, so does the probability of introducing inconsistencies in the resulting model. This methodology is easy to use but very heavy in real life situations.

3   Approach

Our approach is to reduce the cognitive workload of applying the spreadsheet strategy (and similar ones) by offering supporting tools for each of the steps of the process. The resulting platform is called Logikós. Tools for the management of *web information objects* support the earlier activities while *shared decision profiles* support the later ones. Figure 2 provides an overview of the resulting approach. To model items of interest, the decision makers use the templates editor. The result is a template that indicates how to extract an information object from pages that match a given URL pattern. Templates are shared via the templates repository. The items collector uses the available templates to help the user collect items of interest. Collected items are stored in a private repository in the user's web-browser. Using the Multi Criteria Decision Making tool, the decision maker explores the various applicable decision profiles (available in a shared repository) to explore alternative ranking of the collected items. Decision profiles are created and shared by a community of like-minded users, supported by AHP modeling tools. A community of like-minded users is easy to detect thanks to similarity measures introduced in Social Networks.

The following sections discuss each of these elements in more detail.

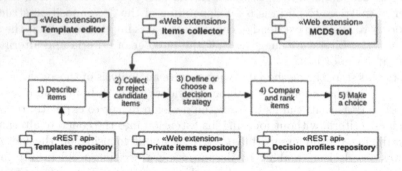

Fig. 2. Approach overview.

3.1   Web Information Objects

This approach draws on the facilities offered by Web Objects Ambient (WOA), a platform for collaborative web augmentation [5]. It helps final users collect, structure and share information available on the web, and wield it to attach new functionality to the sites and applications they visit. For this, they do not require any programming skills.

WOA is as a toolbox. Tools can be combined to support different scenarios. The core tools support extraction and storing of information objects from the web. They are the templates editor, the information objects collector, a repository of templates, and two repositories for the collected information objects (a private one and a shared one).

A template editor tool lets users define templates to collect information objects available on websites. A template has a type (identified by a Semantic Web class resource), a URL pattern, and a list of properties that make up the information object. Objects of the same type can be available in various sites. A template for a given object type tells how to obtain such objects from pages with a particular DOM structure. The URL pattern of the templates tells in which pages the template applies to. Properties have a name and an XPath selector. The selector indicates the specific route in the DOM tree that must be followed to obtain the value of the property.

The information object collector is available in all pages that match the url-Pattern of one of the templates in the templates repository. If several templates are available for a given URL, the user can choose which one to apply to extract an information object. Extracted information objects can be stored either in a private repository (local to the user's machine) or in a shared repository.

WOA is implemented following web-augmentation techniques, which means that the functionality it provides can be attached to any website. Web Augmentation is a set of techniques aimed at providing users a means to adapt Web applications to their requirements [4]. There are multiple ways to achieve Web Augmentation (like bookmarklets or extensions at the client, DOM manipulation at a proxy). WOA relies on a client-side technique: the Document Object Model (DOM) manipulation through the API of webextensions[3].

3.2 Shared Decision Profiles

A decision profile captures how a group of similarly minded users would rank a set of alternatives in terms of preference. It is a decision model based on the properties available on those objects and generates a ranking (from most to less preferable). For example, the "A motor cultivator for home scale farmers" profile captures how family farmers select cultivators, weighing its features such as tilling width, tilling height, transmission type, and price. A similar profile is the one called "A cultivator for farmers cooperative", that while looking at the same properties of the same information object, ranks giving properties different weights.

A shared decision profile is built as a reusable analytic hierarchy process model (all nodes but the alternatives). The model is the result of the collaborative work of representatives of a certain community of users (family farmers, cooperative farmers, etc.). Following a collaborative process and using existing tools (such as Superdesicions[4], or GRUS [9]), users provide the relative weights of the available features of a certain type of item. They then export the model to the repository of decision profiles. The model can be used to obtain the local and overall priorities of a given node. The repository of decision profiles includes a simple, non-collaborative, web-application to create, edit, test, and publish shared decision profiles. A shared decision profile does not depend on the specific alternatives that are the focus of the decision. It only depends on the type of items (in fact, on its properties). Therefore, it can be reused whenever a decision maker understands the profile represents her interests. Only when no available shared profiles refer to the type of item that will be evaluated, or does not represent the interests of the stakeholder, it is necessary to create and share a new one.

3.3 Limiting the Need of Pairwise Comparisons Among Alternatives

Using shared profiles removes the need for pairwise comparisons among criteria. Still, as alternatives depend on the specific situations, pairwise comparisons among alternatives need to be performed. As we have previously discussed, the effort of conducting pairwise comparisons among the alternatives grows with the number of alternatives and criteria. To reduce or even eliminate the need of conducting such comparisons, we introduce the concept of "smart ranking strategy" (or SRS). An SRS describes how pairs of alternatives are to be compared in terms of a given attribute without the need for the decision maker's intervention.

---

[3] https://browserext.github.io/browserext/ - Last accessed on Nov, 2018.
[4] https://www.superdecisions.com/ - Last accessed on Nov, 2018.

The UML class diagram presented in Fig. 3 highlights the main elements of a shared profile using SRS. A shared profile has a title, a description and the types of items it helps choose (the type of items is related to the type of items extracted by templates). The root of the decision tree (the goalModel), and the intermediate decision tree nodes (criteria and sub-criteria) are modeled as instances of class Node. The class Alternative extends Node to add the attributes that were extracted by the template. Criteria nodes that refer to tangible properties of the items (such as price, weight, transmission type, etc.) can have an SRS. Three types of SRS are presented in the model (although it can be extended by adding new subclasses in the future).

- A NormalizedNumericDifferenceSRS compares to alternatives in terms of one of their numeric attributes (e.g., price). It takes the difference of the two values, divides the difference by the largest of them and transforms its absolute value to the scale 1 to 9.
- An EquallySpreadValuesSRS has a list of all the possible values for the given item attribute. Then, to compare two alternatives, it obtains the distance between the positions (in the list of allPossibleValues) of the values of both alternatives. This difference is then divided by the total number of possible values and mapped to Saaty's 1 to 9 scale.
- A StorePairwiseComparisonsSRS stores pairwise comparisons for every pair of elements in the domain of a property, so other decision-makers do not have to perform them. This strategy is only practical when the domain of the property has a manageable number of possible values, and when tools so aid pair-wise comparisons are available (for example, to minimize inconsistency).

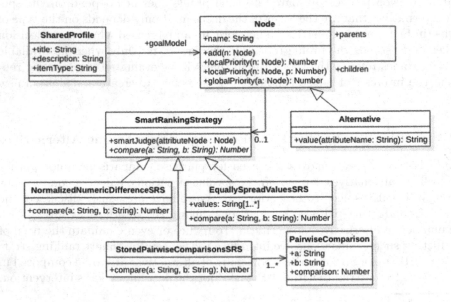

Fig. 3. UML class model of a shared profile

4   Descriptive Evaluation: A Scenario

The following detailed scenario provides evidence of the utility of our approach to reduce the effort of using a MCDM approach in everyday purchase decisions.

Faustino is a farmer managing a family business that is expanding. He has decided to buy new motor-cultivators, expecting that these artifacts will help his employees with the tillage. He went to the local agribusiness and asked for advice. The salesperson offered multiple cultivators but never asked what he wanted the cultivator for. It did not feel right; there are cultivators better suited for certain kind of soil and dimensions to cover (e.g., having in mind depth and width of the work, type of transmission). Last time Faustino bought tools (a smartphone for everyone in the farm), he used an excel spreadsheet that a colleague suggested. When he asked for a similar spreadsheet but for motor-cultivators, his colleague pointed Faustino to the Logikós website instead[5]. These are the steps he followed to finally make a purchase he felt certain with.

Getting started: With some help from the video-tutorials in the "Installing the toolset" section, he was able to install the browser extension and learn how to use it. He first installed the web-extensions and created a new, private, list of alternatives. He didn't install a server for the templates, he used the default shared public repository. This took about the same time as installing the spreadsheet tool. Next time, the tools will be already available.

Collecting items: After a web search, he identified a couple of web-sites that sold cultivators. He started by visiting the molamaq.com web-site. Molamaq.com is a site for motor-based agricultural implements. He realized that another users had already defined a template for cultivators in that site by following the steps 1,2,3 and 4 of Fig. 4. The existing template extracts the name of the tool, the price, the width, height and transmission form. Faustino browsed the site looking for cultivators and, as a template was found, the Items Collector was already prepared to extract items from the site. Faustino extracted eight cultivators from this site that he considered within his budget. He did it as shown in steps 6 and 7 of the Fig. 5. At this point, he realized the difference of this approach. There was not need to copy and paste all attributes and the URL of the page. He did not make any of the common copy&paste mistakes he did last time (e.g., pasting to a wrong row or column). He felt like he would collect as many candidates as he wanted.

Defining a template: Then, he visited mercadolibre.com.ar, a general-purpose e-commerce. He found a cultivator he likes, but he was not able to collect it. Logikós was not prepared to collect items on this website. But following the steps mentioned in Fig. 4, Faustino defined an extraction template and push it to the public repository for other users to use. The template extracted the same attributes as the template he used for Molamaq. He collected other six

---

[5] https://sites.google.com/view/logikos/demo-videos.

cultivators from Mercadolibre. He had already collected two of those (the same brand and model) from the first website he visited. However, they were cheaper in mercadolibre.com.ar. Ok, this was something really new. It did not take much time, but he has to learn it. When he finished he realized why the previous step was so simple; someone had prepared a template. He also realized he was now helping others. A new template for this site will not be needed as long as the structure of the site does not change.

Fig. 4. Creating a template for "Molamaq" cultivators

Exploring the alternatives list: In step 8 of Fig. 6, Faustino opened his list of collected alternatives through the "Private Item's repository" and realized he had collected 16 candidates (for 12 different cultivators) from 2 different online agribusinesses. This part felt exactly as it was with the spreadsheet. He could sort, filter, and browse.

Multi-criteria decision making: Faustino explored different ranking strategies presented as "Decision Profiles", as you can appreciate in step 9 of Fig. 6. Such profiles were prepared by agronomists and farmers who were proficient in the use of AHP. Choosing a strategy re-sorts the list of candidates according to a function weighting the different features of the cultivators. There was, for example, a profile labeled "Home farmer" weighing price and transmission as the two most important criteria (0.39 and 0.35 respectively). Tilling width and tilling height are less important (0.17 and 0.09 respectively). Faustino learned this by

Fig. 5. Extracting items by using the "Molamaq" template

looking at the pie chart diagram offered by the tool. Moreover, when comparing alternatives concerning price, tilling width, and tilling height the profile uses a strategy called "normalized difference" (basically using the numerical values of those properties to decide which one of them is better and for how much). According to that strategy, a cultivator that costs $14000 is "extremely better" than one that costs $38000, but only "moderately better" that one priced at $17.000. The transmission property can have only one of three values "gears, chain, belt". The shared profile offers a "stored pairwise comparison" for this property. Compared to the spreadsheet he used last time, this approach was more transparent. The descriptions and figures that accompanied each profile gave him a better idea of the criteria used to rank (this was not possible with the spreadsheet). Moreover, he could explore other strategies to make the same decision (the spreadsheet had only one).

Decision profiles: When Faustino used Logikós there were three different decision profiles available for cultivators. These profiles were created by groups of users that wanted to help others make informed decisions. Each profile includes a graphical description (a pie chart) of the relative weights given to the item properties. They also include a textual description to help users decided on the applicability of the profile to them (or their needs). As the profile has been authored by a group of volunteer users, access to the group's discussion forum is available. Faustino felt like he could be part of such community, helping them better understand what the needs of a farmer like him and his neighbors.

5   Related Work

A key innovation of our approach is supporting the collection. Closely related work has been done by Quinn and colleagues [6]. They explored the challenges of data collection for decision makers that were proficient in the use of spreadsheets. First, they approach data gathering as a (declarative) crowd-sourcing task. The decision maker indicates the pieces of data that are needed (queries in a spreadsheet cell), and other users will try to find the data. In this sense, it

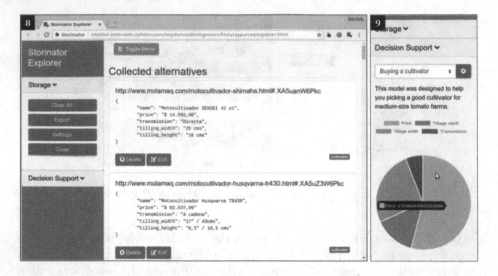

Fig. 6. Evaluating the alternatives with a decision profile

differs from our approach (using a shared repository) in that users do not ask others to find information, but share information they find. In order to reduce the effort of data collection, the authors propose to focus on those pieces of data (spreadsheet cells) that will have the most impact on a decision. In our approach, as the decision maker can explore various decision profiles, data needs to be available in advance. However, a hybrid approach could be explored in the future.

Our approach (based on WOA) relies on being able to robustly abstract content from web-pages. Although it can be seen as a form of content scrapping it does not intend to cover the same goals. Extraction templates are intended to be applied by users to collect only those pieces of information they find valuable. Still, some scraping techniques could be applied to merge information about the same object obtained from various sources, or to keep the information up-to-date. A similar connection could be made principles of the Semantic Web [2]. Our goal is not to create a semantic version of a web-page so it can be interpreted by computers but to obtain information that can be used specifically in MCD. Moreover, it as the semantic web grows, we will tap into its power to complement the information collected by users. In fact, we are gradually incorporating some of its concepts (classes and attributes) to describe the structure of collected items.

6   Conclusions

Literature identifies four key steps in cognitive decision making: (1) Intelligence - collecting information; (2) Conception - creating a model for the decision; (3) Choice - using the model to select the best alternatives; and

(4) Review - coming back to the intelligence step to have more information, redesign the model, and to chose again. It is an iterative process and non-linear. We argue that this process is poorly supported in the web, especially in the context of e-commerce. In this article we presented Logikós, an approach to augment any web-site with support for MCDA. Logikós supports the decision maker during the intelligence, choice, and review phases. It help users to make decisions in a personalized way by choosing the decision profile that best matches their needs. Besides, it conceives the modeling step as a collaborative activity, giving users that are proficient with MCDA the chance to create and share reusable decision profiles. Less-experienced users can then apply such profiles, completing the cognitive decision making cycle. One perspective of this work can be to combine our approach with others MCDM methods like SWING, SMART etc.

Acknowledgement. The authors of this publication acknowledge the support of the Project 691249, RUC-APS: Enhancing and implementing Knowledge based ICT solutions within high Risk and Uncertain Conditions for Agriculture Production Systems (www.ruc-aps.eu), funded by the European Union under their funding scheme H2020-MSCA-RISE-2015.

References

1. Akarte, M., Surendra, N., Ravi, B., Rangaraj, N.: Web based casting supplier evaluation using analytical hierarchy process. J. Oper. Res. Soc. **52**(5), 511–522 (2001)
2. Antoniou, G., Harmelen, F.V.: A Semantic Web Primer, Cooperative Information Systems, vol. 57. The MIT Press, Cambridge (2004)
3. Brown, R.: Consideration of the origin of Herbert Simon's theory of "satisficing" (1933-1947). Manag. Decis. **42**(10), 1240–1256 (2004)
4. Díaz, O.: Understanding web augmentation. In: Grossniklaus, M., Wimmer, M. (eds.) ICWE 2012. LNCS, vol. 7703, pp. 79–80. Springer, Heidelberg (2012). https://doi.org/10.1007/978-3-642-35623-0_8
5. Firmenich, S., Bosetti, G., Rossi, G., Winckler, M., Barbieri, T.: Abstracting and structuring web contents for supporting personal web experiences. In: Bozzon, A., Cudre-Maroux, P., Pautasso, C. (eds.) ICWE 2016. LNCS, vol. 9671, pp. 77–95. Springer, Cham (2016). https://doi.org/10.1007/978-3-319-38791-8_5
6. Quinn, A.J., Bederson, B.B.: Asksheet: efficient human computation for decision making with spreadsheets. In: Proceedings of the 17th ACM Conference on Computer Supported Cooperative Work and Social Computing - CSCW 2014 (2014)
7. de FSM Russo, R., Camanho, R.: Criteria in AHP a systematic review of literature. Procedia Comput. Sci. **55**, 1123–1132 (2015)
8. Saaty, T.L.: Decision making with the analytic hierarchy process. Int. J. Serv. Sci. **1**, 83–98 (2008)
9. Zaraté, P., Kilgour, D.M., Hipel, K.: Private or common criteria in a multi-criteria group decision support system: an experiment. In: Yuizono, T., Ogata, H., Hoppe, U., Vassileva, J. (eds.) CRIWG 2016. LNCS, vol. 9848, pp. 1–12. Springer, Cham (2016). https://doi.org/10.1007/978-3-319-44799-5_1

Dynamic-R: A New "Convincing" Multiple Criteria Method for Rating Problem Statements

Oussama Raboun[1](✉), Eric Chojnacki[2], and Alexis Tsoukias[1]

[1] Univ. Paris-Dauphine, PSL Research University, CNRS, UMR [7243], LAMSADE, 75016 Paris, France
oussama.raboun@dauphine.eu, tsoukias@lamsade.dauphine.fr
[2] Institut de Radioprotection et de Sureté Nucléaire (IRSN), Cadarache, France
eric.chojnacki@irsn.fr

Abstract. In this paper, we propose Dynamic-R method, a new decision aiding procedure to deal with multicriteria rating problems. A multicriteria rating problem consists on partitioning a set of objects, assessed under several dimensions called criteria, into predefined ordered equivalence classes, called categories, identified by rates. Several rating methods were developed using the majority rule. These methods present many disadvantages leading potentially to an unconvincing rating. In this work, we introduce a dynamic rating procedure aiming at providing a "convincing" rating (stable under criticisms) over a set of studied objects. It is called dynamic, since rated objects will be used to characterize the categories in the next steps. The developed rating procedure is based on the aggregation of positive and negative reasons respectively supporting and opposing to a rating.

Keywords: Multicriteria decision aiding ·
Rating problem statements · Decision support systems ·
Algorithmic decision theory

1 Introduction

In this paper we propose a new MCDA (Multiple Criteria Decision Aiding) method for rating problems [6]. This work can be seen as a particular case of ordinal classification problems, consisting in partitioning a set of objects, here after named A, into predefined ordered equivalence classes, called categories. Since we are dealing with objects evaluated under several dimensions, called criteria, we will consider rating problem statements in the context of MCDA.

Several MCDA methods have been developed to deal with rating problems. These methods can be partitioned into three categories: [i.] methods based on the majority principle, called *outranking* methods (e.g. [1,2,10,17,19,20]); [ii.] methods based on the assessment of utility functions (e.g. [5,8,9,14,16]); and

© Springer Nature Switzerland AG 2019
P. S. A. Freitas et al. (Eds.): EmC-ICDSST 2019, LNBIP 348, pp. 136–149, 2019.
https://doi.org/10.1007/978-3-030-18819-1_11

[iii.] methods based on rough sets (e.g. [7,12,13,15]). In this work, we are interested to the same type of problems for which outranking methods fit.

Outranking methods, in the context of rating problems, are based on preference relations established between the set A and reference profiles without considering comparisons among objects. This property can be seen as an extension of the IIA (independence of irrelevant alternatives) property. Because of this property, outranking methods may lead to non-convincing ratings, either because of the non existence of any preference structure preventing from cycles among objects assigned to different categories (Condorcet Paradox), or because of incomparabilities. This is because Outranking relations do not have any remarkable properties, see [3]. Consider the following example:

Example 1. (Non convincing rating due to Condorcet Paradox)
Let us consider a rating problem characterized by three necessary and sufficient criteria, i.e. the three are exhaustive and none of them is a dictator. This comes to considering any coalition of two criteria as a decisive coalition. We consider that each criterion evaluates the set A on an ordinal scale: $\{B, A, A^+\}$. In this problem we aim at assigning two objects $x = (A^+, A, B)$ and $y = (A, B, A^+)$ into two predefined ordered categories C_1 (rate 1) and C_2 (rate 2) such that C_1 is the best. The two categories are separated by a lower bound of C_1: $p = (B, A^+, A)$. Using the majority rule to rate x and y, we obtain: $y \succ p$ and $p \succ x$, where $\succ$ refers to the strict preference relation. Thus, y will be rated 1 while x will be rated 2. However $x \succ y$ with a 2/3 majority.

The originality of this work consists on proposing a new "dynamic" and "convincing" rating MCDA method, named "Dynamic-R", for problems characterized by ordinal information under at least one criterion and without considering the IIA axiom. A multicriteria "convincing" rating is a resulting ordinal classification of elements in A, based on clear positive and negative reasons and without any contradiction. The dynamic aspect of the method is related to the rating procedure associated to the method: the rated objects are added to the profiles characterizing the categories and are used in the next step when new objects are consider for rating. In order to obtain a "convincing" rating, and in order to use the rated objects as reference profiles characterizing the categories (the dynamic aspect), we address the following properties:

- We allow comparison among elements in the set A;
- We allow both limiting and typical profiles;
- We separate positive and negative reasons for and against a rating;
- We provide a monotonic and complete rating, as a result of our rating procedure.

The paper is organized as follows. Section 2, introduces notations used all along the paper. In Sect. 3, we discuss some existing methods and introduce the developed. In Sect. 4, we present the concepts and the properties characterizing the developed method. We end by a conclusion and a discussion.

2 Notations and Concepts

All along this document we will use the following notation:

- A set of steps $I = \{1, 2, 3, ...\}$. These indices represent steps in which a new set of objects is considered for a rating.
- At each step $i \in I$, a new set of studied objects $A^i = \{x, y, z, w, ...\}$ is considered for rating, called alternatives or actions [4]. Here after we will use A^i instead of A.
- In this paper we will use the mathematical notation "$[\![\,;\,]\!]$", to refer to the integer interval.
- A set of predefined ordered categories $\mathcal{C} = \{\mathcal{C}_1, ..., \mathcal{C}_q\}$, $q \geq 2$, where $\mathcal{C}_k$ refers to a category where objects are rated k. Without loss of generality, we assume that $\mathcal{C}_1$ is the best category, and $\forall k \in [\![1\,;\,q-1]\!]$ $\mathcal{C}_k$ is better than $\mathcal{C}_{k+1}$.
- Criteria $\mathcal{F} = \{1, ..., m\}$ with $m \geq 3$ evaluating the studied objects. Each criterion $\mu \in \mathcal{F}$ is associated to a weak order $\succcurlyeq_\mu$ on the set A^i.
- Importance of criteria w, is a capacity defined as: $w : 2^{\mathcal{F}} \to [0, 1]$. ($w(\mathcal{F}) = 1$, $w(\emptyset) = 0$, and for all $A, B \in 2^{\mathcal{F}}$ such that $A \subseteq B$, $w(A) \leq w(B)$).
- Importance of the discordant criteria $\mathcal{V}$, is a capacity defined as: $\mathcal{V} : 2^{\mathcal{F}} \to [0, 1]$. By definition of capacity we have $\mathcal{V}(\mathcal{F}) = 1$ (all criteria reject a given preference), $\mathcal{V}(\emptyset) = 0$, and for all $A, B \in 2^{\mathcal{F}}$ such that $A \subseteq B$, $\mathcal{V}(A) \leq \mathcal{V}(B)$.
- Reference profiles, at a step $i \in I$, $Z^i = \{Z_1^i, ..., Z_q^i\}$, where $Z_h^i = \{z_{h,k}, k = 1, ..., t_{h,i}\}$, $t_{h,i} \geq 1$, represents the set of reference profiles characterizing the category $\mathcal{C}_h$, at the step i. The initial set of reference profiles Z^0 is used as a learning set to generate the preferential information. We will use also the notation: $\forall j, k \in [\![1\,,\,q]\!], j < q : Z_{j,k}^i$ to refer to $\underset{t \in [\![j\,,\,k]\!]}{\cup} Z_t^i$.
- A set of minimal requirements $\mathcal{B} = \{b_1, ..., b_q\}$, characterizing categories where performances of the profile $b_k = (b_{j,k})_{j \in \mathcal{F}}$ characterizing $\mathcal{C}_k$, are the minimal performances to be admissible in $\mathcal{C}_k$. These minimum requirements are characterized by the following condition: We assume that $\forall j \in \mathcal{F}, \forall h \in [\![1\,;\,q-1]\!] : b_h \succ_j b_{h+1}$. To not confuse b_k with a limiting profile since b_k does not belong necessarily to $\mathcal{C}_k$.
- A set of all objects $\mathcal{A}^i = A^i \cup_k Z_k^i \cup \mathcal{B}$ considered at the step i of the rating aggregation procedure.
- Parameters: λ the sufficient considered majority, called concordance threshold; v the veto threshold.
- Negative reasons against being rated t or higher, based on the comparison with reference profiles: $\forall x \in A^i \forall t \in [\![1\,;\,q-1]\!] : x R_r^- Z_t^i$.
- Positive reasons for being rated t or lower, based on the comparison with reference profiles: $\forall x \in A^i \forall t \in [\![1\,;\,q]\!], x R_r^+ Z_t^i$.
- Withdrawn negative reasons against being rated t, due to comparisons among objects: $\forall x \in A^i \forall t \in [\![1\,;\,q-2]\!] : x R_{rr}^- Z_t^i$.
- Enriched positive reasons for being rated t, due to comparisons among objects $\forall x \in A^i \forall t \in [\![1\,;\,q-1]\!] : x R_{ur}^+ Z_t^i$.
- The set of objects for which the lowest possible rating is k, with respect to reference profiles and objects in A^i: L_k^i.

- The set of objects for which the highest possible rating is k, with respect to reference profiles and objects in A^i: H_k^i.
- Positive reasons for being rated t, due to a minimization of the distance between objects in $A^i \setminus \cup_k \left(H_k^i \cap L_k^i \right)$ and the reference profiles: R_{2r}^+.

3   Overview of Dynamic-R

In this section, we describe Dynamic-R. First, we present the general architecture of the existing rating methods based on the majority rule and their extensions with a consistency checking. Then we provide a description of the Dynamic-R and problems to which it fits. Finally we address the rating procedure.

3.1   Outranking Methodology for Rating Problems

The existing rating procedures based upon the use of outranking relations use a majority principle applied on positive reasons, this being bounded by a minority principle (usually a veto condition) which can invalidate the aggregation of the positive reasons. Positive reasons are typically obtained comparing objects either to limiting profiles separating categories, or to typical profiles characterising the categories. In the first case we make use of asymmetric comparisons, while in the second case we make use of symmetric comparisons. Objects are never compared among them.

Several rating methods have been developed aiming at rating a set of objects with respect to a consistency rule. For example, Rocha and Dias in [18] developed a PASA (Progressive Assisted Sorting Algorithm) algorithm, respecting the following consistency principle: an object cannot be assigned to a category in case it is outranked by any example (reference profile) assigned to a lower category. This principle seems very close to our work since we also characterize the categories by a set of reference profiles and we have a consistency rule. However, this method presents also many disadvantages such as:

- the order of the selected objects for rating might bias the ratings of the next selected objects;
- in case of an imprecise rating, either the decision maker is needed or the rating is postponed;
- forcing the consistency might lead to bad quality of rating: objects involved in cycles are placed in the same category (the lower category among the ones to which objects can be assigned).

THESEUS method [11] is an other rating method, aiming at providing a rating minimizing inconsistencies with respect to a learning set (reference profiles in our case). This method is based on an original approach, transforming a rating problem into a ranking problem. Such transformation consists on associating to each non rated object x, new alternatives x_k: "assign x to the category k". The generated alternatives x_k are assessed under the following criteria: inconsistencies with respect to the strict preference, the weak preference, and the indifference. Hence, the problem of rating x, comes to a ranking problem associated to selecting the best x_k, minimizing the inconsistencies. We address the following weaknesses of THESEUS method:

- The provided rating minimizes inconsistencies. However, it does not provide a convincing rating;
- The dependency on the learning set: both small and very big learning sets may lead to a poor rating either because of incomparabilities or the high number of inconsistencies.

The next section will be dedicated to present Dynamic-R and its advantages.

3.2   Presentation of Dynamic-R

Dynamic-R is a MCDA rating method based on the majority rule and discrimination through the use of minimum requirements. In our work, the minimum requirements are profiles representing the minimum acceptable performances, under each criterion regardless the global performance, to be in a category. For instance, regardless the global mark, a student cannot be considered a good student if he performs less than 7/20 in any of the lectures. Minimum requirements should not be confused with limiting profiles: in the previous example $(7/20, ..., 7/20)$ is the minimum requirement associated to the category of good students, however, a student performing 7/20 in all the lectures "$(7/20, ..., 7/20)$" is a bad student. Dynamic-R introduces four new ideas:

1. positive and negative reasons are kept separated, and we only consider at the last step how to aggregate them;
2. does not make any distinction between limiting and typical profiles since both of them might be available and provide positive or negative reasons about the rating of a given object x;
3. explicitly introduces the concept of minimal requirements, a disjunctive constraint among the criteria, providing strong evidence that an object CANNOT be rated to a certain category (because it fails to satisfy a requirement on any of the criteria), without the vector of minimal requirements being a profile of any category;
4. it cumulates reference profiles since objects that are rated at step i are used as profiles both for step $i + 1$, but also as consistency checking within step i, thus allowing comparisons among objects.

Dynamic-R can be seen as a method learning from the history of ratings done by a decision maker, it makes this history convincing by updating all inconsistencies (this represents the reference profiles sets in Z^0), and automatize the rating process. The rating process starts by a Z^0 respecting the convincing condition: "No better reference profile assigned to a lower category". Each step i of the process is characterized by a new set of objects to be rated A^i. The rating process associated to Dynamic-R, at a step i, can be structured as follow:

1. For each object in $x \in A^i$, we compute positive and negative reasons, for and against rating x to a category $\mathcal{C}_k$ for all k (based on the way they compare to reference profiles $R^+_{1,r}$ and R^-_r).
2. We revise positive and negative reasons for each object, and reference profile based on the way objects compare to each other (R^+_{ur} and R^-_{rr}).
3. We compute H^i_k and L^i_k, $\forall k \in [\![1 \,;\, q]\!]$, All objects in $H^i_k \cap L^i_k$ will be assigned to Z^{i+1}_k. We distinguish two cases:
   (a) Objects belonging to any among the sets $H^i_1 \cap L^i_1, ..., H^i_q \cap L^i_q$. In other terms, objects having the same maximum and minimum rating. These objects are rated k.
   (b) Objects not belonging to $\cup_k H^i_k \cap L^i_k$. Such objects have different minimum and maximum rating, we can consider them as interval rated. In such case, we compute a distance between objects and reference profiles characterizing the possible categories and we choose the "nearest" one. This is done through the use of R^+_{2r}. The distance is computed first over objects in H^i_1, then H^i_2,..., and we end by objects in H^i_q. Each time an object is rated based on the distance we revise positive reasons for the other objects (return to step 2.)

We have to mention that the resulting rating of procedure presented above, is used in the next step $i + 1$, when new objects are considered for rating. This is due to enriching the set of reference profiles in $i + 1$ by the rating of the step i.

The reader should note that Dynamic-R is a whole rating process, rather a simple rating procedure. Under such a perspective the "convincing" property of Dynamic-R refers to the outcome of the whole process.

4   Basic Concepts Within Dynamic-R

Dynamic-R is a MCDA rating method based on defining and aggregating positive and negative reasons respectively for and against a rating. The concepts used in our method will be presented in this section.

4.1   Basic Definitions

Definition 1. *(Positive reasons for an outranking)*
   Positive reasons for outranking relations are binary relations R^+ defined on $(\mathcal{A}^i)^2$ representing the capacity of a sufficient coalition of criteria to influence the relative preference between two objects. This can be expressed as:

$$xR^+y \iff w(\{j \in \mathcal{F} : x \succeq_j y\}) \geq \lambda \tag{1}$$

where λ is the majority threshold

Definition 2. *(Weak dominance relation)*
   A weak dominance relation D is a binary relation defined on $(\mathcal{A}^i)^2$. For $x, y \in \mathcal{A}^i$, we say that xDy if x is at least as good as y under each criterion and strictly better than y under at least one criterion. This can be formulated by:

$$xDy \iff \exists i \in \mathcal{F}, \forall j \in \mathcal{F} : x \succeq_j y \wedge x \succ_i y \tag{2}$$

In this paper, many definitions involve binary relations between objects and the sets of reference profiles. We propose the following formulation:

Definition 3. *(Binary relations used in positive and negative reasons)*
Consider the set A and a set of sets B. A binary relation $\mathcal{R} \subseteq A \times B$, such that $\forall (x, Y) \in A \times B : x\mathcal{R}Y$ should be read as "there are positive reasons for x to belong to Y", or "there are negative reasons for x belonging to Y".

In assignment problems where categories are not necessarily ordered, the assignment is based on similarity indices which can be seen as a distance between an object and reference profiles characterizing a class.

Definition 4. *(Distance between an object and a set of characteristic profiles)*
Let Z_k^i be a set of reference profiles characterizing $\mathcal{C}_k$ at a step $i \in I$. The distance of $x \in A^i$ from the set Z_k^i, $dist(x, Z_k^i)$, can be formulated as:

$$dist(x, Z_k^i) = \min \left(\min_{z \in Z_k^i} |c(x, z) - c(z, x)|; \frac{1}{|Z_k^i|} \left| \sum_{z \in Z_k^i} c(x, z) - c(z, x) \right| \right) \quad (3)$$

where $c(x, y) = w(\{j \in \mathcal{F} : x \succeq_j y\})$

This distance computes the minimum between the closest reference profile characterizing a category to an object based on the difference of criteria importance's in favor of the object and the reference profile, and how the object is located compared to all reference profiles characterizing a category. The first component of the distance, $\min_{z \in Z_k^i} |c(x, z) - c(z, x)|$, represents the minimum of distances between "x" and each profile in Z_k^i. Intuitively, it can be seen as an answer to the question "is there any profile in Z_k^i close to x?". The second component of the distance, $\frac{1}{|Z_k^i|} |\sum_{z \in Z_k^i} c(x, z) - c(z, x)|$, represents the net flow evaluation: The difference between total importance of criteria in favor of x compared to the profiles in Z_k^i and the total importance of criteria in favor of the reference profiles in Z_k^i compared to x. This last can be seen as an evaluation of the distance with the center of reference profiles in Z_k^i.

We defined an incompatibility binary relation between categories and objects based on the minimum requirements.

Definition 5. *(Incompatibility binary relation)*
Incompatibility binary relation $Incomp_{lower}$ defined on $A^i \times Z^i$, represents the non illegibility of an object to characterize a given category with respect to some minimum requirements. For $x \in A^i$, $Z_k^i \in Z^i$:

$$x Incomp_{lower} Z_k^i \iff \neg(x D b_k) \text{ such that } b_k \in \mathcal{B} \quad (4)$$

4.2 Theoretical Foundations of Dynamic-R

In this section, we present the theoretical foundations of Dynamic-R. We define different types of positive and negative reasons as well as their properties.

Negative reasons represent information or premises against a classification. In our approach negative reasons against rating k, an object x, represent the situation where an x is either weakly dominated by a reference profile characterizing $k+1$ or incompatible with k.

Definition 6. *(Negative Reasons against rating)*
Negative reasons against a rating are binary relations R_r^- defined on $A^i \times Z^i$. For $x \in A^i, Z_k^i \in Z^i$, $x R_r^- Z_k^i$ can be formulated as:

$$x R_r^- Z_k^i \iff (\exists h > k, \exists z \in Z_h^i : zDx) \vee x Incomp_{lower} Z_k^i. \tag{5}$$

If Definition 8 holds then:

Proposition 1. *(properties of negative reasons)*

1. If there exist negative reasons against assigning an object to a given category then there exist negative reasons against assigning it to any better category:

$$\forall x \in A^i, \forall Z_h^i \in Z^i : x R_r^- Z_h^i \implies \forall k \leq h : x R_r^- Z_k^i; \tag{6}$$

2. If there exists no negative reason to assign an object to a given category then there exists no negative reasons to assign it to any worse category:

$$\forall x \in A^i, \forall Z_h^i \in Z^i : \neg(x R_r^- Z_h^i) \implies \forall k \geq h : \neg(x R_r^- Z_k^i). \tag{7}$$

Positive reasons are built on preferential information supporting a rating. Such information is based on the monotonicity of the assignment. Positive reasons consists on the existence a sufficient majority of criteria in favor of an object in comparison with at least a profile characterizing a higher category.

Definition 7. *(Positive reasons supporting a rating)*
For $i \in I$, positive reasons R_{1r}^+ are binary relations defined on $A^i \times Z^i$ representing the possibility to be at least as good as reference profiles characterizing a category. For $x \in A^i, Z_k^i \in Z^i$, R_{1r}^+ can be formulated by:

$$x R_{1r}^+ Z_k^i \iff \exists h \leq k, \exists z \in Z_h^i, x R^+ z. \tag{8}$$

The set of characteristic profiles must respect the following basic "convincing" condition:

Definition 8. *"Convincing" condition can be formulated as:*

$$\forall z \in Z_k^i, \nexists y \in Z_h^i(h > k), y R^+ z \wedge \neg(y R_r^- Z_k^i) \tag{9}$$

A peculiar feature of the proposed method is that at each step i, the set of reference profiles is updated since the last rated objects are added and used for the step $i + 1$. The consequence is that "positive" and "negative" reasons computed at step i need to be updated at step $i + 1$.

Example 2. Let us consider the example 1, such that, at a given step i, $y = (A^+, A, B)$ and $p = (B, A^+, A)$ are reference profiles characterizing respectively the second category (the lowest) and the highest category. We aim at rating a new object $x = (A, B, A^+)$ and we suppose that x is compatible with the highest category. Since $\neg(yDx)$, then $\neg(xR_r^- Z_1^i)$. Also since $xR^+ p$ then $xR_{1,r}^+ Z_1^i$. However, this rating cannot be inconvincing unless $y = (A^+, A, B)$ is not compatible with the highest category. The same situation would happen in case we aim to rate x and y at the same step, since they were not compared during the assessment of positive and negative reasons, we might have an inconvincing result.

Keeping in mind that our method is expected to maintain the "convincing" property, and it is necessary at step $i + 1$ to perform (hierarchically) three updates: enrich positive reasons, withdraw negative reasons, which we present in the following.

Remark 1. (Notation)
To simplify notation we will use the following notations:

- $Z_{j,k}^i$ to refer to $\underset{t \in \{j,\dots,k\}}{\cup} Z_t^i$;
- $\forall t \in [\![1\,;\,q]\!], U_{r,t}^- = \{w \in B^i \cup Z_{1,q}^i, wR_r^- Z_t^i\}$;
- $\forall t \in [\![1\,;\,q]\!], U_{r,t}^+ = \{w \in B^i \cup Z_{1,q}^i, wR_{1r}^+ Z_t^i\}$.

Definition 9. *($U_{ur,k}^+$, For $k \in [\![1\,;\,q-1]\!]$)*

For a given $k \in [\![1\,;\,q-1]\!]$, the set of objects, $U_{ur,k}^+$, for which positive reasons were enriched to support a rating k, can be formulated as:

$$U_{ur,k}^+ = \{w \in Z_{1,q}^i \cup A^i : wR_{ur}^+ Z_k^i \wedge \neg(wR_{ur}^+ Z_{k-1}^i)\} \tag{10}$$

where R_{ur}^+ is binary relation representing enriched positive reasons supporting a rating. R_{ur}^+ can be formulated as: For $x \in U_{r,t}^+ \setminus U_{r,t-1}^+$:

$$xR_{ur}^+ Z_k^i \iff \exists y \in (\cup_{j=1}^k U_{ur,j}^+ \cup U_{r,k}^+) \setminus U_{r,k}^- : xR^+ y \tag{11}$$

Definition 9, represents the assessment of the sets of objects for which positive reasons were enriched to support the assignment to a higher category. Enriching positive reasons for a given x is mainly due to the presence of y, having positive and no negative reasons to be assigned to a category better than x, such that $xR^+ y$. Hence, y will provide x by new positive reasons that will potentially improves its possible rating.

The following proposition presents a characteristic of the binary relation used in the assessments of $U_{ur,1}^+, \dots, U_{ur,q}^+$.

Proposition 2. *(properties of R_{ur}^+)*

For $i \in I$, for $x \in U_{r,t}^+ \setminus U_{r,t-1}^+$ we have the following property:

$$xR_{ur}^+ Z_k^i \implies \forall j \in [\![k\,;\,t-1]\!] : xR_{ur}^+ Z_j^i \tag{12}$$

Definition 10. ($U_{rr,k}^-$, For $k \in [\![1\,;q-1]\!]$)

For a given $k \in [\![1\,;q-1]\!]$, the set of objects, $U_{rr,k}^-$, for which negative reasons were withdrawn to prevent a higher rating k, can be formulated as:

$$U_{rr,j}^- = \{w \in Z_{1,q}^i \cup A^i : wR_{rr}^- Z_j^i \wedge \neg(wR_{rr}^- Z_{j+1}^i)\} \tag{13}$$

where R_{rr}^- is binary relation representing withdrawn negative reasons against a rating. R_{rr}^- can be formulated as: For $x \in U_{r,t}^- \setminus U_{r,t+1}^-$:

$$xR_{rr}^- Z_k^i \iff \begin{cases} \forall z \in (U_{r,t}^-) \setminus \cup_{j=1}^{t-1} U_{rr,j}^- & : \neg(zDx) \\ \text{And} \\ \exists y \in [\cup_{j=1}^{t} U_{ur,j}^+ \cup U_{r,t}^+] \cap N_{t,k}^- & : yDx \vee xIncomp_{lower} Z_k^i \end{cases} \tag{14}$$

with $N_{t,k}^- = \cup_{j=k}^{t-1} U_{rr,j}^- \cup U_{r,k}^- \setminus (\cup_{j=1}^{k-1} U_{rr,j}^- \cup U_{r,t}^-)$ representing objects with valid negative reasons against ratings between $t-1$ and k.

Definition 10, represents the assessment of the sets of objects for which negative reasons were withdrawn to prevent a rating to a higher category. The binary relation R_{rr}^- associated to these sets (and characterizing the operation of withdrawing negative reasons for a given x from a lower rating l, to a higher rating h) is defined by two conditions:

1. Eligibility for withdrawing negative reasons: no object or reference profile, having valid negative reasons against the rating l, dominates x. Otherwise, x will still have valid negative reasons against being rated l.
2. New negative reasons against a rating k: the improvement of the rating of x will be at most limited by the improvement of the object or reference profile, let's name it y, at the origin of x's negative reasons. The limitation might also come from an other element dominating x, limiting its improvement to at most $k+1$. It is also possible that the withdrawn of x's negative reasons will not be limited by any object or reference profile, but by its own performance not dominating the minimum requirement b_k.

Proposition 3. (properties of R_{rr}^-)

For $i \in I$, $\forall x \in U_{er,t}^- \cup (U_{r,t}^- \setminus U_{r,t+1}^-)$, we have the following property:

$$xR_{rr}^- Z_k^i \implies \forall j \in [\![1\,;k]\!] : xR_{rr}^- Z_j^i \tag{15}$$

Remark 2. The updates of negative and positive reasons led to a change of some reference profiles to a higher category. The updated sets of reference profiles can be formulated as:

$$Z_{uk}^i \doteq Z_k^i \setminus \left[(\cup_{j=1}^{k-1} U_{ur,j}^+ \cap (\cup_{j=1}^{k-2} U_{rr,j}^-)) \cup (\cup_{j=1}^{k-1} U_{ur,j}^+ \setminus U_{r,k-1}^-) \right] \tag{16}$$

Since the sets of reference profiles are updated, we have more objects to rate. Thus, we note A_u^i the new set of objects that need to be rated at the current step $i \in I$. A_u^i can be formulated as:

$$A_u^i = A^i \cup \left(Z_{1,q}^i \setminus (\cup_{j=1}^{q-1} Z_{uj}^i) \right) \tag{17}$$

The assessments of the $U^+_{r,t}$, $U^+_{ur,t}$, $U^-_{r,t}$, and $U^-_{rr,t}$, for all t, need to be used in order to have a "convincing" rating. For this aim, A^i_u will be partitioned into $H^i_1, ..., H^i_q$, and $L^i_1, ..., L^i_q$. These partitions will be defined, based on a binary relation between A^i_u and Z^i_{uk}, for all k, as follows:

Definition 11. *(H^i_h and L^i_l, for $h, l \in [\![1\,;q]\!]$)*

For a given $i \in I$, the partitions of A^i_u, H^i_h and L^i_l, for which the highest and the lowest possible ratings are respectively $h, l \in [\![1\,;q]\!]$, can be formulated as:

$$\begin{cases} H^i_h = \{x \in A^i_u, Z^i_{u,h} \succcurlyeq^i x \land \neg(Z^i_{u,h+1} \succcurlyeq^i x)\} & h \neq q \\ H^i_q = \{x \in A^i_u, Z^i_{u,q} \succcurlyeq^i x\} \end{cases} \tag{18}$$

$$\begin{cases} L^i_l = \{x \in A^i_u, x \succcurlyeq^i Z^i_{u,l} \land \neg(x \succcurlyeq^i Z^i_{u,l-1})\} & l \neq 1 \\ L^i_1 = \{x \in A^i_u, x \succcurlyeq^i Z^i_{u,1}\} \end{cases} \tag{19}$$

Where $\succcurlyeq^i$ being a weak order built on $(A^i_u \times Z^i_u) \cup (Z^i_u \times A^i_u)$, representing the preference between a subset of A^i_u and sets in Z^i_u. $\forall i \in I, \succcurlyeq^i$, is defined as follow:

1. On $Z^i_u \times A^i_u$: $\forall x \in A^i_u, \exists k \in [\![1\,;q]\!]$ such that:

$$Z^i_{u,k} \succcurlyeq^i x \iff \begin{cases} x \in U^-_{r,k-1} \setminus \cup^{k-2}_{j=1} U^-_{rr,j} & k > 2; \\ x \in U^-_{r,k-1} & k \leq 2; \end{cases} \tag{20}$$

2. On $A^i_u \times Z^i_u$: $\forall x \in A^i_u, \exists k \in [\![1\,;q]\!]$ such that:

$$x \succcurlyeq^i Z^i_{u,k} \iff \begin{cases} \neg(Z^i_{u,k+1} \succcurlyeq^i \{x\}) \land (x \in \cup^k_{j=1} U^+_{ur,j} \cup U^+_{r,k}) & k \neq q \\ x \in \cup^q_{j=1} U^+_{ur,j} \cup U^+_{r,q} \end{cases} \tag{21}$$

With these elements we can establish a first rating. This rating concerns objects for which the highest and lowest possible rating lead to the same category: objects in $H^i_k \cap L^i_k$. However, the rating of objects is not always precise: there exist objects for which the best possible category and the worst possible category are not the same, objects in $A^i_u \setminus (\cup^q_{k=1} H^i_k \cap L^i_k)$. Such objects require additional information in order to be rated. This information can be seen as additional positive reasons supporting a rating to one of the categories located between the highest and the lowest possible categories. For this aim, we define a symmetric binary relation based in the distance function *dist*, see Definition 4. This function represents a similarity measure evaluating how close is an object from an updated set of reference profiles Z^i_u.

Definition 12. *($U^+_{2r,k}$, for $k \in [\![1\,;q]\!]$)*

$U^+_{2r,k}$, for $k \in [\![1\,;q]\!]$, refers to the set of objects for which the rating is not precise and the closest updated reference profiles are the ones rated k. For $k \in [\![1\,;q]\!]$, $U^+_{2r,k}$ can be formulated as:

$$U^+_{2r,k} = \{w \in \left((\cup^k_{j=1} H^i_j) \cap (\cup^q_{j=k} L^i_j)\right) \setminus (H^i_k \cap L^i_k); x R^+_{2r} Z^i_{u,k}\} \tag{22}$$

where R_{2r}^+ is a binary relation defined on $A_u^i \times Z_u^i$, that can be interpreted for $(x, Z_{u,k}^i)$ as "x is as good as reference profiles characterizing C_k". For $x \in A_u^i$, R_{2r}^+ can be formulated as:

$$x R_{2r}^+ Z_{u,k}^i \implies Z_{u,k}^i = \arg \min_{Z \in K_x \subseteq Z_u^i} dist(x, Z) \tag{23}$$

where $K_x = \{Z_{u,k}^i \in Z_u^i; x \in \left((\cup_{j=1}^k H_j^i) \cap (\cup_{j=k}^q L_j^i) \right) \setminus (H_k^i \cap L_k^i) \}$.

K_x consists on sets of reference profiles characterizing categories for which the rating of the object x is not precise based on $\succsim^i$.

Definition 12 represents the assessment of the objects having a second level of positive reasons. This last represents additional reasons supporting a rating. These reasons can be interpreted as the capability of an object to describe a category based on how close it is to the sets of reference profiles. The second level of positive reasons might also provide additional positive reasons for other objects: each object assigned based on the second level of positive reasons may enrich positive reasons for objects in lower categories and more precisely the ones for which the assignment is not precise. Hence, it is possible to not assess the second level of positive reasons for all objects for which the assignment is not precise. Hence, the order of assessing the second level of positive reasons is very important.

5   Discussion and Conclusion

This paper presents a new method that provides a "convincing" rating to MCDA problems with at least one ordinal dimension. It takes into consideration the way objects compare to each other. The method yields a dynamic rating of objects through an aggregation of reasons for and against a rating. The main idea behind the method is the learning from an evolving set of reference profiles characterizing each category. A step is characterized by new objects considered for rating which terminates with an assignment of the rated objects to the corresponding sets of reference profiles. Many perspectives might be associated to the developed method, such as: the importance of criteria might change during the process; the possibility to assess positive and negative reasons on coalitions of objects such as the case of an insurance company aiming at rating a package of clients or products that might interact; to name but a few. A specific mention should be given to the development of an argumentation framework for explaining/justifying/defending a rating thanks to the explicit representation of the positive and negative reasons on which such rating has been established.

References

1. Almeida-Dias, J., Figueira, J., Roy, B.: Electre Tri-C: a multiple criteria sorting method based on characteristic reference actions. Eur. J. Oper. Res. **204**(3), 565–580 (2010)
2. Almeida-Dias, J., Figueira, J., Roy, B.: A multiple criteria sorting method where each category is characterized by several reference actions: the electre Tri-NC method. Eur. J. Oper. Res. **217**(3), 567–579 (2012)
3. Bouyssou, D.: Outranking relations: do they have special properties? J. Multicriteria Decis. Anal. **5**(2), 99–111 (1996)
4. Bouyssou, D., Marchant, T., Pirlot, M., Tsoukiàs, A., Vincke, P.: Evaluation and Decision Models with Multiple Criteria: Stepping Stones for the Analyst, 1st edn. Springer, Boston (2006). https://doi.org/10.1007/0-387-31099-1
5. Bugera, V., Konno, H., Uryasev, S.: Credit cards scoring with quadratic utility functions. J. Multi-criteria Decis. Anal. **11**(4–5), 197–211 (2002)
6. Colorni, A., Tsoukiàs, A.: What is a decision problem? Preliminary statements. In: Perny, P., Pirlot, M., Tsoukiàs, A. (eds.) ADT 2013. LNCS (LNAI), vol. 8176, pp. 139–153. Springer, Heidelberg (2013). https://doi.org/10.1007/978-3-642-41575-3_11
7. Dembczyński, K., Greco, S., Słowiński, R.: Rough set approach to multiple criteria classification with imprecise evaluations and assignments. Eur. J. Oper. Res. **198**(2), 626–636 (2009)
8. Dembczyński, K., Kotłowski, W., Słowiński, R.: Additive preference model with piecewise linear components resulting from dominance-based rough set approximations. In: Rutkowski, L., Tadeusiewicz, R., Zadeh, L.A., Żurada, J.M. (eds.) ICAISC 2006. LNCS (LNAI), vol. 4029, pp. 499–508. Springer, Heidelberg (2006). https://doi.org/10.1007/11785231_53
9. Devaud, J.M., Groussaud, G., Jacquet-Lagreze, E.: UTADIS: Une methode de construction de fonctions d'utilite additives rendant compte de jugements globaux. In: European Working Group on MCDA, Bochum, Germany (1980)
10. Fernández, E., Figueira, J., Navarro, J., Roy, B.: Electre Tri-NB: a new multiple criteria ordinal classification method. Eur. J. Oper. Res. **263**(1), 214–224 (2017)
11. Fernandez, E., Navarro, J.: A new approach to multi-criteria sorting based on fuzzy outranking relations: the theseus method. Eur. J. Oper. Res. **213**(2), 405–413 (2011)
12. Greco, S., Matarazzo, B., Slowinski, R.: Rough approximation by dominance relations. Int. J. Intell. Syst. **17**(2), 153–171 (2002)
13. Greco, S., Matarazzo, B., Slowinski, R.: Rough sets methodology for sorting problems in presence of multiple attributes and criteria. Eur. J. Oper. Res. **138**(2), 247–259 (2002)
14. Greco, S., Mousseau, V., Słowiński, R.: Multiple criteria sorting with a set of additive value functions. Eur. J. Oper. Res. **207**(3), 1455–1470 (2010)
15. Greco, S., Matarazzo, B., Slowinski, R.: Rough sets theory for multicriteria decision analysis. Eur. J. Oper. Res. **129**(1), 1–47 (2001)
16. Köksalan, M., Özpeynirci, S.B.: An interactive sorting method for additive utility functions. Comput. Oper. Res. **36**(9), 2565–2572 (2009)
17. Leroy, A., Mousseau, V., Pirlot, M.: Learning the parameters of a multiple criteria sorting method. In: Brafman, R.I., Roberts, F.S., Tsoukiàs, A. (eds.) ADT 2011. LNCS (LNAI), vol. 6992, pp. 219–233. Springer, Heidelberg (2011). https://doi.org/10.1007/978-3-642-24873-3_17

18. Rocha, C., Dias, L.C.: An algorithm for ordinal sorting based on electre with categories defined by examples. J. Glob. Optim. **42**(2), 255–277 (2008)
19. Vincke, Ph.: Outranking approach. In: Gal, T., Stewart, T., Hanne, T. (eds.) Multicriteria Decision Making, Advances in MCDM Models, Algorithms, Theory and Applications, pp. 11.1–11.29. Kluwer Academic Publishers, Dordrcht (1999)
20. Wei, Y.: Aide multicritère à la décision dans le cadre de la problématique du tri: Concepts, méthodes et applications. Ph.D. thesis, Paris 9 (1992)

Author Index

Printed in the United States
By Bookmasters